TELECOMMUNICATIONS TECHNOLOGY

R. L. BREWSTER, Ph.D, MSc
Lecturer in Telecommunications
University of Aston in Birmingham

ELLIS HORWOOD LIMITED
Publishers · Chichester

Halsted Press: a division of
JOHN WILEY & SONS
New York · Chichester · Brisbane · Toronto

First published in 1986 by
ELLIS HORWOOD LIMITED
Market Cross House, Cooper Street, Chichester, West Sussex, PO19 1EB,
England

*The publisher's colophon is reproduced from James Gillison's drawing of the ancient Market
Cross, Chichester.*

Distributors:

Australia and New Zealand:
Jacaranda-Wiley Ltd., Jacaranda Press,
JOHN WILEY & SONS INC.
GPO Box 859, Brisbane, Queensland 4001, Australia

Canada:
JOHN WILEY & SONS CANADA LIMITED
22 Worcester Road, Rexdale, Ontario, Canada

Europe and Africa:
JOHN WILEY & SONS LIMITED
Baffins Lane, Chichester, West Sussex, England

North and South America and the rest of the world:
Halsted Press: a division of
JOHN WILEY & SONS
605 Third Avenue, New York, NY 10158, USA

© **1986 R. L. Brewster/Ellis Horwood Limited**

British Library Cataloguing in Publication Data
Brewster, R. L.
Telecommunications technology. — (Ellis Horwood series in electrical and communications
engineering)
1. Telecommunication — Technological innovations
I. Title
631.38   TK5102.5

Library of Congress Card No. 85–27347

ISBN 0–85312–908–8 (Ellis Horwood Limited)
ISBN 0–470–20270–X (Halsted Press)

Typeset by Ellis Horwood Limited
Printed in Great Britain by R. J. Acfords, Chichester

TELECOMMUNICATIONS
TECHNOLOGY

Table of Contents

Preface

The art of telecommunications is passing through an exciting stage of development and in the years between now and the turn of the century we are likely to see many new developments which will significantly affect our whole way of life. There have been a number of recent developments in telecommunications technology which have brought considerable change to the existing communications network, with many of the traditional modes of communication now fading into obsolescence. This book has been written to bridge the gap between the old and the new and thus the emphasis is rather different from that normally expected in a basic book on the telecommunications network. For this reason, and also for the very good reason that it is my own specialist interest, the reader will find rather more on digital transmission and networks and rather less on the conventional, but fast-disappearing, analogue network that has served us well for almost a century. However, it will be a long time yet before all the old plant is dispensed with and in the mean time the telecommunications engineer must take cognisance of its existence. I hope the reader will feel that sufficient has been included to meet this need and that the proportion of space allocated to its consideration is roughly proportional to its significance as part of the network, at least for the immediate future.

It is the purpose of this book to give a broad, yet fairly in-depth, introduction for undergraduates studying telecommunications as a specialist option, and a general review of telecommunications for those already qualified in engineering but are desirous of broadening their knowledge into the field of telecommunications, perhaps with a view to a change of discipline into a rapidly expanding growth area of modern technology.

The book has been prepared from notes of lectures given mainly as a course in Telecommunications to second- and third-year undergraduates at Aston University. The notes have been gathered from many sources and frequently updated over a number of years. The origins of much of the information are therefore no longer known and it would be impossible to cite references in any acceptable way. I have, therefore, rather included a bibliography at the end of the book which includes those books I have on my shelf which have been an inspiration to me over the years I have taught the course, those books I have from time to time included in the recommended reading lists, and those books I regard as classic works on specific topics and to which I turn for more specialised information. I hope the reader will find the list useful.

I suspect that several colleagues past and present may find some features in the book vaguely familiar. I have not intentionally copied from other sources, though diagrams especially are sometimes held in memory with some precision, particularly those that have proved to be of greatest help to me in my own study of the subject. I therefore hope that any colleague finding something that looks rather familiar will take it as a compliment to his original clarity rather than an act of blatant plagirism on may part. I can do no more than thank those many colleagues who over the years have given me helpful support and inspiration, both in teaching and research. The acquisition of knowledge depends on mutual interaction with others and without their help, albeit indirect, this book would not have been possible.

Figures 6.7, 6.12, 6.13, 6.14, 6.21, 6,22, 6.23, 6.27 and 8.1 are reproduced, by kind permission, from *Local Telecommunications,* Ed. J. M. Griffiths, published by Peter Peregrinus.

Birmingham, July 1985   R. L. Brewster

1

Introduction

Communication is the process of exchanging information. Telecommunications is the science of communicating over distances where the basic modes of communication, such as speech and vision, are no longer feasible. The most widely used method of communicating over such distances is by the use of electrical signals, either over cables or through free space using radio waves. Recently, however, the use of light signals carried over optical fibres has become a practical alternative, with considerable implications for the future of telecommunications.

Telecommunications signals can be divided into two classes. The best known, and by far the most widely used method of person-to-person telecommunications, is the telephone. The telephone converts speech pressure waves into electrical signals, which are transmitted over the telephone network and then converted back into sound waves in the telephone earpiece. The electrical signals are complex waveforms which can take on any value. Such signals are known as continuous, or analogue, signals.

A fast-growing area of telecommunications is communication with and between computers. Computers communicate using digital signals which take on predetermined discrete values only. These signals, therefore, differ fundamentally from those used in telephony and thus require different transmission facilities. As will be seen later, modern technology makes the transmission and switching of digital signals a much more flexible and reliable process than that traditionally used for analogue telephone speech signals. It is now becoming common practice, therefore, to convert analogue signals into digital signals, by a process of quantisation, before transmission over a digital network. The quantisation process is known as pulse code modulation (PCM). Pulse code modulation means that telephony and data can now use the same digital network and the current trend is towards the development of an integrated services digital network (ISDN), which will handle all telecommunications services without discrimination.

In the mean time, there is frequently a requirement for data transmission facilities where, at present, no digital network access exists. Since the telephone network is accessible to almost all the inhabited world, it is the obvious potential carrier for data signals that need access to a central computer from remote locations. Techniques have therefore been developed to modulate data signals onto an analogue carrier-wave to make them suitable for transmission over the telephone network. The devices used to interface the data terminals to the analogue network are known as 'modems', an acronym of MOdulator—DEModulator.

The telephone network does not consist simply of pairs of wires and switches. For efficient long-distance (trunk) communication, several signals are combined together, or multiplexed, so that they may be carried simultaneously by a single cable of appropriate quality rather than a large bundle of separate wire pairs. For transmission to geographically inaccessible places, radio channels are used. Over difficult terrain it is often better to use microwave links than to have to lay cables in the ground. For international communication, satellites now offer a more flexible medium than the laying of submarine cables on the ocean bed.

Before we commence an in-depth study of telecommunications technology there are one or two topics with which we need to become familiar. Any reader who is already familiar with the basic principles of communications theory may wish to skip the rest of this chapter and proceed directly to the start of Chapter 2.

A GENERAL MODEL OF THE TELECOMMUNICATIONS PROCESS

It will be of help to us if we can begin with a general model of the telecommunications process. Such a model is given in Fig. 1.1. Firstly we have to obtain an electrical signal which represents the information from our information source. In our general model we have designated this process 'source-to-signal encoding'. In the context of this model the word 'encoding' should be given the widest possible interpretation. Perhaps the commonest of all source-to-signal encoders in use in telecommunications at the present time is the microphone used in the telephone handset. This converts the sound pressure waves into electrical signals whose voltage amplitude varies in sympathy with the sound wave amplitude variations. The electrical signals thus occupy the same frequency band as the sound wave variations. These electrical signals may not necessarily be suitable for direct transmission over the chosen transmission medium. For example, sound wave signals cannot be propagated directly as broadcast radio waves. They therefore have to be translated into the radio frequency spectrum before they can be transmitted as radio waves. The conversion of the signals into a form suitable for the transmission medium is carried out in the transmitter-to-medium

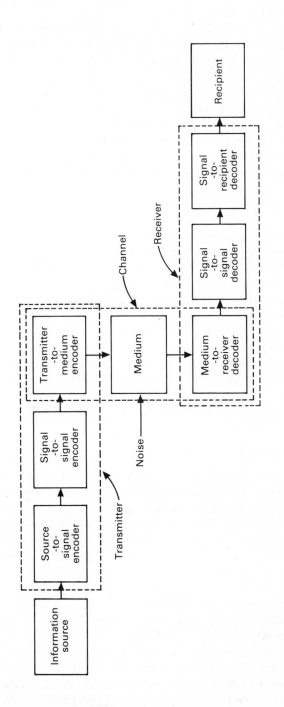

Fig. 1.1 – General model of telecommunications process.

encoder. Often an intermediate stage of signal processing is required to effect the communication process and this is designated in our model as the 'signal-to-signal encoder'. This stage is particularly significant in digital transmission, where error-detecting or correcting code elements may be added to the source signal before it is converted into a line code for onward transmission. The process of converting the source information into a form suitable for the transmission medium is the function normally associated with the transmitter equipment.

A complementary set of decoding operations are performed in the receiver equipment so that information is finally passed on in a form that is intelligible to the recipient. Thus, for example, the electrical signals are converted back into sound pressure waves in the telephone earpiece. In some applications, for example in data transmission and optical fibre transmission, it is convenient to consider the transmission medium, together with the transmission-to-medium encoder and medium-to-receiver decoder, as an entity we shall call the 'communication channel'. Thus there is not always a distinct boundary between the transmitter, channel and receiver elements of the communication process. We shall now look in a little more detail at one of the most basic of all transmitter-to-medium encoding processes – that of modulation.

MODULATION

Modulation is the process whereby signals which naturally occur in a given frequency band, known as the base-band, are translated into another frequency band so that they can be matched to the frequency characteristics of the transmission medium. Thus, for example, electrical signals created by sound waves have to be translated into the radio frequency spectrum before they can be broadcast for radio communication purposes. They then have to be translated back into the base-band, by a complementary process known as demodulation, before they can be used to create signals which are audible to the recipient.

The process of modulation translates the signal from the base-band into a frequency spectrum associated with a 'carrier' frequency. If we chose our carrier to be a sinusoidal waveform of frequency f_c and of arbitrary phase and amplitude, we may define the unmodulated carrier waveform, $C(t)$, as follows:

$$C(t) = A \cos(2\pi f_c t + \phi) .$$

We can now modulate our carrier by varying one of the parameters in accordance with the amplitude of the modulating base-band signal. There are three parameters we could select for modulation. Firstly we could modulate the amplitude, A, of the carrier signal, a process referred to as amplitude modulation.

AMPLITUDE MODULATION

Let us consider a carrier $C(t)$ which is amplitude modulated by a modulating signal $M(t) = B \sin 2\pi f_m t$. Then our modulated signal

$$V(t) = B \sin 2\pi f_m t \cdot \cos 2\pi f_c t$$

$$= B \left(\frac{\sin 2\pi (f_c - f_m)t}{2} + \frac{\sin 2\pi (f_c + f_m)t}{2} \right)$$

Thus our signal consists of two side-band components at frequencies equal to the sum and difference between the carrier and the modulating signal frequencies. (Note we have dropped the phase constant ϕ in the above equation. This would simply displace both side-bands by the same angle ϕ). Note that the carrier frequency has disappeared altogether. In fact it is frequently more convenient to retain a component at the carrier frequency to aid in the demodulation process. If the modulating signal is superimposed on a d.c. signal such that the resultant signal always maintains a positive polarity, then we can define our modulating signal as

$$M(t) = B(1 + m \sin 2\pi f_m t)$$

where $m \leqslant 1$ and is called the modulation index.
    Then

$$V(t) = B(1 + m \sin 2\pi f_m t) \cos 2\pi f_c t$$

$$= B \left(\frac{m \sin 2\pi (f_c - f_m)t}{2} + \cos 2\pi f_c t + \frac{m \sin 2\pi (f_c + f_m)t}{2} \right)$$

Now our signal consists of a pair of side-bands and a carrier component.
    Such a signal is illustrated in Fig. 1.2(a). The advantage of such a signal is that it can be demodulated simply by extracting the 'envelope' of the signal by rectifying and low-pass filtering of the modulated signal. If m is greater than 1, then envelope detection is not possible. However, the 'suppressed' carrier component can be extracted by filtering and used for synchronous detection of the modulated signal. In synchronous detection, the modulated signal is multiplied by the recovered carrier component so that the demodulated signal

$$D(t) = \cos 2\pi f_c t \left(\frac{m \sin 2\pi(f_c - f_m)t}{2} + \right.$$

$$\left. \cos 2\pi f_c t + \frac{m \sin 2\pi(f_c + f_m)t}{2} \right).$$

(We have dropped the coefficient B since it is only a scaling factor.) This will yield a signal such that each of the three frequency components represented by

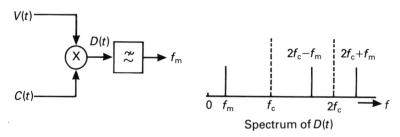

(a) Amplitude modulated signal

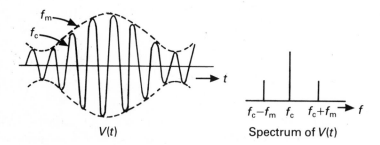

(b) Synchronous demodulation of AM

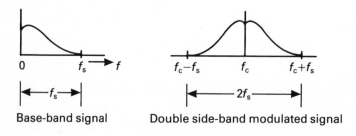

Base-band signal   Double side-band modulated signal

Fig. 1.2 — Amplitude modulated signal spectra.

the terms in the large brackets form a pair of sidebands about the carrier frequency f_c. These components are illustrated in Fig. 1.2(b). It will be noted that the lowest frequency component is in fact the modulating signal itself. Thus the modulating signal can be recovered by simple low-pass filtering. If the modulating signal is a complex signal consisting of a band of frequency components, then each individual component will produce a side-band pair. The bandwidth of the amplitude modulated signal is thus twice that of the modulating signal, as illustrated in Fig. 1.2(c). From an information theory point of view, both side-bands in this 'double side-band' signal carry the same information. One of the side-bands is therefore redundant and can, in theory, be dispensed with. We can therefore conserve transmission bandwidth by removing one of the side-bands and transmitting a 'single side-band' signal. In principle, it does not matter which of the two side-bands is retained. However, synchronous demodulation is necessary and therefore some knowledge of the carrier frequency is required by the demodulator. Since the two side-bands may be situated in close proximity to the carrier, it is not always possible to totally remove one side-band without also removing the carrier component. In such a case it is sometimes preferable to maintain a vestige of the other side-band in a technique known as vestigial side-band modulation. We shall not discuss vestigial side-band modulation in detail here as it is of specialised interest only. The interested reader is referred to the many excellent books on modulation theory for more details, if required.

FREQUENCY MODULATION

Instead of varying the amplitude A of the carrier in sympathy with the modulating signal, it is possible to vary the frequency f_c. Thus

$$V(t) = \cos 2\pi f_c (1 + B \sin 2\pi f_m t)t .$$

This may be rewritten in the form

$$V(t) = \cos(2\pi f_c t + \beta \sin 2\pi f_m t) \tag{1.1}$$

where $\beta = \Delta f / f_b$ and is known as the modulation index. Δf is the maximum frequency deviation of the carrier frequency and f_b is the maximum bandwidth of the base-band modulating signal. For a sinusoidal modulating signal, $f_b = f_m$. The bandwidth of the modulated signal is seen to be related both to the frequency deviation Δf and to the amplitude B of the modulating signal.

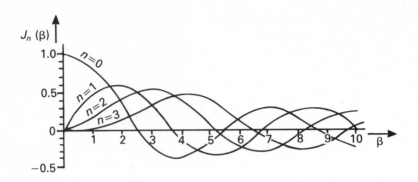

Fig. 1.3 – Bessel functions of the first kind, $n = 0$ to 3.

Expanding equation (1.1) gives

$$V(t) = \cos 2\pi f_c t \cos(\beta \sin 2\pi f_m t) - \sin 2\pi f_c t \sin(\beta \sin 2\pi f_m t)$$

$$(1.2)$$

For small values of β ($\beta < 0.2$) we can use as an approximation to this equation

$$V(t) = \cos 2\pi f_c t - \beta \sin 2\pi f_m t \cos 2\pi f_c t .$$

This is similar to the equation for amplitude modulation and gives rise to a single pair of side-bands and a carrier component, the side-bands being at the sum and difference frequencies of the carrier and the modulating signal.

For values of $\beta > 0.2$ we need to expand equation (1.2) further and to do this we need to make use of a set of mathematical functions known as the 'Bessel functions of the first kind'. These are denoted by the symbol $J_n(\beta)$. The Bessel functions are normally given in tabular form. A plot of the first few Bessel functions ($n = 0$ to 3) is given in Fig. 1.3.

In terms of the Bessel functions, equation (1.2) becomes

$$
\begin{aligned}
V(t) = &\ J_0(\beta) \cos 2\pi f_c t - J_1(\beta) \left[\cos 2\pi (f_c - f_m)t - \cos 2\pi (f_c + f_m)t\right] \\
&+ J_2(\beta) \left[\cos 2\pi (f_c - 2f_m)t + \cos 2\pi (f_c + 2f_m)t\right] \\
&- J_3(\beta) \ \cos 2\pi (f_c - 3f_m)t - \cos 2\pi (f_c + 3f_m)t] \\
&+ \dots \text{etc.}
\end{aligned}
$$

Thus the modulated signal is comprised of a carrier component and side-bands which in principle extend to infinity at frequencies spaced at $\pm f_m, \pm 2f_m, \pm 3f_m$ etc. about the carrier frequency. The significance of the Bessel functions of higher order decreases rapidly with decreasing β. As a rough guide, the number of significant side-band pairs can be taken to be approximately equal to $\beta + 2$

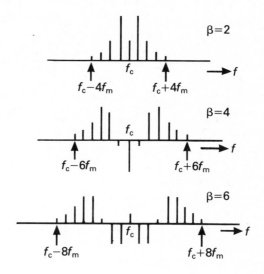

Fig. 1.4 – Spectra of FM signals for various β.

for all reasonable values of β. Some examples of signal spectra are given in Fig. 1.4. Thus the bandwidth required to transmit a frequency modulated signal increases roughly in proportion to the modulation index (β), whereas for amplitude modulation the bandwidth required is independent of the modulation index (m). However, frequency modulation (FM) has the advantage that the amplitude of the modulated signal contains no information whatsoever regarding the modulating signal. Thus channel attenuation and, in particular, fading have no directly harmful effect on the overall transmission of signals through the communication channel.

PHASE MODULATION

The final alternative to varying the amplitude A or the frequency f_c of the carrier is to vary the absolute phase ϕ of the carrier signal. In fact, since a continuous variation in phase is directly equivalent to a variation in frequency, for a smoothly varying analogue signal there is no fundamental difference between frequency and phase modulation. Effectively phase modulation is frequency modulation where the modulation is carried out using the derivative of the modulating signal. Under these conditions the spectrum of the modulated signal is similar to that obtained using frequency modulation. However, with a digital modulating signal which changes instantaneously between discrete amplitude levels, the derivative of the modulating signal is quite different in nature from the original signal. Under these circumstances, frequency modulation becomes a series of instantaneous changes in carrier frequency, usually referred to as

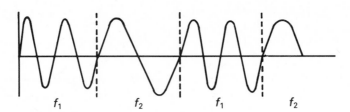

f_1 f_2 f_1 f_2

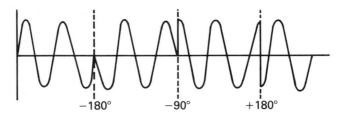

$-180°$ $-90°$ $+180°$

frequency-shift-keying (FSK). On the other hand, with phase modulation there is no change in the carrier frequency, simply steps in the signal phase, usually referred to as phase-shift-keying (PSK). Examples of FSK and PSK are given in Fig. 1.5. Thus for digital modulation the similarity between frequency and phase modulation disappears and different modulated signal spectra are obtained. Digital modulation, including FSK and PSK, is discussed in detail in Chapter 6.

THE DECIBEL

Before we leave this introduction it will be as well if we acquaint ourselves with the use of the decibel as a unit of measure in telecommunications systems. Actually the decibel is not an absolute unit of measure at all, it is in fact a convenient way of specifying power ratios.

A communication channel may consist of a number of cascaded stages, some of which may be amplifiers and others of which may be attenuators. Both gain and attenuation may be specified as a power ratio in the form

$$G_p = \frac{\text{Power output}}{\text{Power input}} .$$

It is also possible to specify the gain or attenuation in terms of a voltage ratio

$$G_{\mathrm{v}} = \frac{\text{Voltage at output}}{\text{Voltage at input}} \; .$$

Often attenuation is quoted as the reciprocal of the value given by the above formulae so that both gain and attenuation are given values of greater than unity.

If the input impedance and the load impedance are equal and equal to R_{L}, then

$$\text{Power output} = (\text{voltage at output})^2 \, R_{\mathrm{L}}$$

and

$$\text{Power input} = (\text{voltage at input})^2 \, R_{\mathrm{L}}.$$

Thus

$$G_{\mathrm{p}} = G_{\mathrm{v}}^2 \; .$$

If channel stages are cascaded, then the overall gain is the product of the gains of the individual stages. Often, and especially in the case of amplifying stages, the overall gain can become very large. The human response to increased power, for example the response of the ear to sound signals, is based on a logarithmic rule. Thus equal increments in the response to signals are caused by a multiplicative increase in the actual signal power. It is thus quite convenient to express the signal ratios in logarithmic units. Thus $\log_e G_{\mathrm{p}}$ gives the power ratio in natural logarithmic units, known as nepers (after Lord Napier). In fact it is more common in telecommunications to take the logarithm to base 10. However, the resulting unit, the bel, is too small for practical use. The basic logarithm is thus multiplied by a factor ot 10 to give the unit of the decibel. Thus power ratio in decibels (dB) is given by

$$10 \log_{10} G_p \; \mathrm{dB}.$$

Note this is a power ratio. Sometimes decibels are used for voltage ratios, when strictly the input and load impedances should be equal. Then

$$10 \log_{10} G_p = 10 \log_{10} G_v^2$$
$$= 20 \log_{10} G_v \; \mathrm{dB}.$$

Thus for voltage ratios the multiplying factor becomes 20.

We have already noted that when amplifiers and attenuators are cascaded the basic ratios accumulate on a multiplicative basis. Thus the gains specified in decibels accumulate on an additive basis. Thus, for example, three amplifiers each with a power gain of 100 have a combined gain in cascade of 100 $\times$ 100 $\times$ 100 = 1,000,000. In decibel units a power gain of 100 = 10 $\log_{10}$ 100 = 20 dB. Thus the overall gain = 3 $\times$ 20 = 60 dB. By dividing 60 by 10 and taking the antilogarithm of the result, it can easily be shown that 60 dB does indeed correspond to a power ratio of 1,000,000. Note for an attentuation of 100, equivalent to a gain of 1/100, we get 10 $\times$ $\log_{10}$ (1/100) = $-$ 20 dB and that an attenuation of 1,000,000, equivalent to a gain of 1/1,000,000 = $-$60 dB. Thus attenuation is denoted by a negative sign before the decibel ratio. For very large values of gain or attenuation the decibel ratios are much more manageable than the equivalent absolute ratio figures. The proper use of polarity signs on the decibel ratios for amplifiers and attenuators avoids the anomalies associated with simple ratios. The overall performance of a channel comprising both amplification and attentuation can be calculated by simple addition, taking full cognisance of the polarity signs in the addition process.

Occasionally we find communications engineers using the decibel as an absolute unit of power. When used in this way, the power of the signal concerned is given as a decibel ratio of some standard of power such as 1 watt or 1 mW. In the first case the units are specified as dBW and in the second case as dBm. Thus 1 W = 0 dBm, 1 kW = 30 dBW or 60 dBm and 1 mW = $-$30 dBW or 0 dBm. Note that individual powers cannot be added arithmetically in decibel units. They must first be converted back into watts before direct addition can be properly carried out.

Equipped with these few basic principles we are now ready to look at telecommunication networks in more detail.

2

The telephone network

The oldest, and by far the most widely used, telecommunications facility at the present time is telephony. We shall therefore commence our study with a look at how the telephone network has evolved to its present state and what significant developments have occurred on the way. This will lead us into a study of telephone traffic and how the network is arranged to give the user an acceptable grade of service.

The invention of the telephone in 1876 is usually attributed to Alexander Graham Bell. In his experiments on early telegraph systems he discovered, quite by accident, that speech pressure waves could be converted into electrical signals and back again using an electromagnetic armature vibrating in sympathy with the sound waves. The first telephones to be used were directly connected together over individual circuit wires. However, as more users began to acquire telephones, it became necessary to arrange some method of interconnecting them according to the user's requirements. Thus grew the concept of the telephone 'office' or 'exchange'. In the earliest exchange, which served only a local community, each telephone was connected to a central office and an operator was alerted by a lamp which was made to glow when the handset was lifted from the telephone instrument. The operator then connected her set to the line, found out who the calling party wished to speak to, and connected the two 'subscribers' together by means of a jumper lead. On completion of the call, she removed the jumper lead and waited for the next call from one of the users. As the size of the network grew, it became necessary for separate offices to be established at different locations and for these offices to be interconnected with 'junction' cables. Calls to another area now required the services of two operators, one at each local exchange. As distances increased, sometimes calls would be connected through several offices in tandem. Now several operators could be involved in a single cell. Also, as the number of subscribers on the local exchange increased, it became necessary for the work to be shared by more than one operator and the concept of switching multiples was born.

It was inevitable that, before long, the reliance on human operators would lead to a suspicion or malpractice and victimisation. It was in just such circumstances that a certain Kansas City undertaker, Alman B. Strowger, suspecting he was suffering a loss of trade, proposed an automatic switching system which has been the basis of telephone switching for almost 70 years. This system, known as step-by-step switching, has only recently been superseded by more modern technology, based on cross-point switches, which we shall consider later. Although the step-by-step switching system is now gradually becoming obsolete, it still represents the major proportion of switching equipment worldwide. Not only that, it is a useful concept around which to discuss many of the problems associated with switching systems and to derive some of the most useful aspects of telephone traffic theory. We shall therefore take a closer look at the principles of switching proposed by Strowger, and still often referred to by his name.

STEP-BY-STEP SWITCHING

The basic component in the step-by-step switching system is the so-called 'two-motion' switch. This enables a connection to be established through any one of a hundred sets of switch contacts arranged on the surface of a cylinder in rows and columns as shown in Fig. 2.1. Selection of the the appropriate output is performed in two stages. In the first stage a contact arm is moved upwards vertically by a relay-operated ratchet arrangement, until it is in line with the required row. In the next stage the contact arm is rotated horizontally until it corresponds with the desired outlet contacts. The stepping movement of the contact arm is governed by a series of electrical impulses applied to the relay mechanism which operates an appropriate stepping ratchet. In a very simple system these impulses are obtained directly from the telephone dial. The dial mechanism is rotated manually to select the required digit. After selection of the appropriate digit, the dial returns to its rest position under the control of a speed governor. As it does so, a cam on the dial operates a contact which generates a series of electrical pulses according to the digit dialled. Thus, in a very simple system, it is possible, by dialling two digits, to select any one of a hundred other telephones, numbered 00 to 99.

In practice, of course, it is quite impracticable to fully interconnect every one of the millions of telephone subscribers throughout the world using a unique switch for each subscriber. Instead, common switching equipment is provided which can be called into use as required. Let us first of all consider an exchange

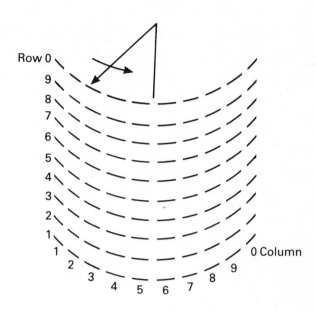

Fig. 2.1 – Strowger two-motion switch contacts.

which would serve 1000 subscribers. The schematic diagram for such an exchange is given in Fig. 2.2. Each subscriber is connected to a single rotary pre-selector switch at the exchange, the outputs from this switch being connected to a bank of two-motion switches known as 'group selectors'. The outputs from the pre-selector switches of a whole group of subscribers are connected together in parallel as that group of subscribers shares a single bank of group selectors. When a subscriber lifts his telephone, the cradle switch causes a

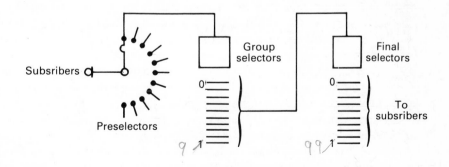

Fig. 2.2 – 1000 line exchange.

circuit to be completed back to the telephone exchange, signalling that the subscriber wishes to make a call. This causes the pre-selector switch to step around until it finds a free group selector. The pre-selector switch then stops in this position and the group selector is 'seized' by the subscriber wishing to make a call. This selector is then no longer available for any other subscriber until the call is released. If, therefore, another subscriber in the group attempts to make a call while the selector is seized, the pre-selector of this subscriber will hunt over this position and continue until it finds a free selector. In the event that all the selectors are in use, engaged tone is returned via the last contact on the pre-selector. The subscriber must then abort his attempt to establish a call and try again later. The art of good switching engineering design is to select the size of the subscriber group and the number of switches in the bank so as to ensure satisfactory service; that is, a small probability only of finding no selector available, at the same time minimising the overall number of selectors that need to be provided.

On seizing the group selector, the subscriber dials his first digit and the selector switch moves up to the appropriate row on the switch contact array. Each final selector has the possibility of connection to 100 lines. The 1000 lines are therefore divided into 10 groups of 100 each, the group being identified by the first digit in the subscriber's number. The vertical motion of the group selector therefore selects a final selector in the group associated with the first digit dialled. Each individual row of contacts, or levels, of the group selector is connected to a bank of final selectors associated with a particular group of 100 line numbers. Having dialled the first digit to select the appropriate group, the group selector arm then automatically rotates in the vertical direction until it finds a free final selector. In the final selector, both directions of motion are under the control of the subscriber's dial and, after dialling two further digits, connection is established, providing the called subscriber's telephone is free at the time. Ringing tone is then connected to signal the called subscriber to answer his telephone. The provision of final selectors is based on similar considerations to those used for group selectors. This arrangement can be extended to form a 10,000 line exchange as illustrated in Fig. 2.3.

By adding further group selector stages it is possible to extend the size of the network almost without limit. It is not, of course, necessary that all stages of the switching need be within a single building. In fact, on a public telephone network, this would only be the case for calls to subscribers connected to the same local exchange. A common method of extending step-by-step switching to a distant exchange is to reserve one or more levels on the first group selector for calls to subscribers on other exchanges. This is often the top level, obtained by dialling the digit '0', which corresponds to ten switching impulses. Further group selectors attached to these levels can then direct the calls to lines connected to the distant exchange. Further stages then finally connect with the final selectors of the distant exchange and thence to the distant subscribers. The call can thus be routed by dialling the appropriate sequence of digits. Calls can, if necessary, be routed via an intermediate exchange; in fact some exchanges in the network

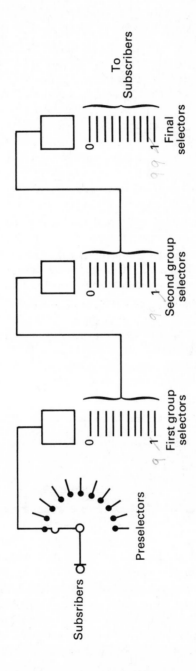

Fig. 2.3. — 10,000 line exchange.

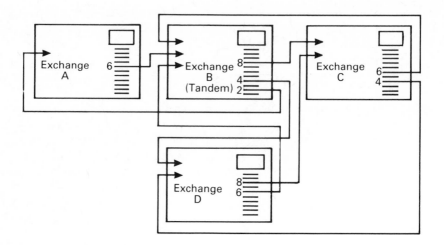

Fig. 2.4 – Basic exchange interconnections.

may only act as intermediaries. Such exchanges, which have no individual subscribers attached, are known as tandems. A schematic diagram of such a switching system is given in Fig. 2.4. The dialling codes associated with this arrangement are shown in Fig. 2.5.

One of the major disadvantages of the simple step-by-step operation just described is that a different sequence of digits has to be dialled to reach the same subscribers, depending on where the call originates from. In a very simple network, such as that shown in Fig. 2.5, it is possible, by careful choice of numbering to avoid this problem. However, the network does not have to be-

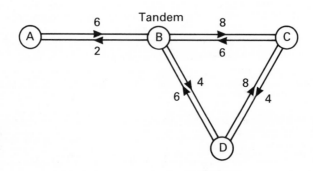

Fig. 2.5 – Dialling codes for Fig. 2.4.

come very complicated before the problem becomes significant. Try devising a numbering scheme to interconnect 9 or 10 exchanges and the problem will become immediately apparent.

SUBSCRIBER TRUNK DIALLING

Modern computing techniques have made possible a fairly simple solution to this problem. In this solution, each subscriber is given a unique number, based on his geographical location and his subscriber number on his local exchange. This number, the subscriber trunk dialling (STD) code, refers only to the subscriber's instrument and does not contain information regarding subsequent routing of the call through the network. When a subscriber wishes to call another subscriber, he now dials the called party's STD number which is stored in a register at the local exchange. Using a code directory stored in memory, the exchange then translates the code stored in the register into the appropriate routing digits. The exchange then generates a sequence of switching impulses in accordance with these digits. These impulses then operate the step-by-step switches just as if they had been dialled by the calling subscriber in the first instance. Nowadays, virtually the whole international telephone network operates on an STD basis, with international dialling codes to specify the destination country of the call being instituted. The full code, including the international pre-fix, is known as the ISTD code. Thus to call Geneva from the UK one would dial:

010	41	22	XXXXXX
International prefix	Code for Switzerland (country code)	Code for Geneva (exchange code)	Subscriber code

For calls to subscribers on one's own local exchange, only the subscriber code digits need to be dialled. For subscribers on other exchanges within the country of origin, the exchange code must precede the subscriber code digits. In the UK the exchange code is often preceded by a '0', which is in fact simply an indication to the exchange that the call is an STD call as opposed to a local call. In this case '0' does not form part of the code when the exchange is called internationally and should not be dialled. For calls to subscribers in other countries, the international prefix (010 in the UK) is dialled, followed by the country code for the destination country. This is then followed by the exchange code and subscriber code as before.

The major disadvantage of the Strowger two-motion switch is its dependence on moving parts and contacts which are subject to wear and tear. This means that the switches need regular maintenance and adjustment, which in turn means they must be sited in locations which are easily and speedily accessible. As the telephone network spread into remote areas, it became necessary to leave exchanges unattended for longer periods and sometimes to site them in locations which are inaccessible during certain seasons of the year. The need was, therefore, for a switching mechanism which would replace the original Strowger switch and would require less maintenance and virtually no re-adjustment after installation. From this need arose the family of switching units known as cross-point switches. Modern electronics technology has made available a whole new range of switching techniques which are ideally suited for cross-point switching systems. However, the first cross-point switching matrix was developed too early to make use of this technology and was therefore still based on a mechanical design.

CROSS-POINT SWITCHING

A diagrammatic representation of a cross-point switching matrix is given in Fig. 2.6. A connection between a particular input terminal and a particular output terminal can be attained by closing the switch contacts at the appropriate cross-point in the matrix. It is then possible, in fact, to establish further connections in this way until there are no more free inlets or outlets from the matrix. The earliest cross-point switching mechanism was that known as 'crossbar', in which metallic contacts at the cross-point were moved together by the simultaneous operation of rotating bars across the matrix in each direction.

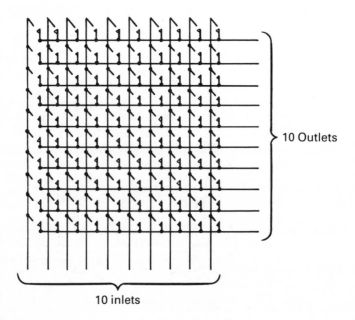

10 Outlets

10 inlets

Fig. 2.6 – Basic cross-point switching matrix.

Since the plane in which the contacts are mounted is usually held vertically we can refer to the cross-bar arms as vertical and horizontal respectively.

The rotation of the horizontal bar selects a row of contacts and rotation of the vertical bar then holds the contacts closed at the selected cross-point. The horizontal selection bar can then be released so that selection of a further contact set can be made. The contacts selected are held closed until the appropriate vertical hold bar is released. To reduce the amount of hardware required, the horizontal selection bars are usually operated by a pair of electromagnets which enables the bar to rotate in either direction, each direction selecting a different row of contacts.

Although the cross-bar switch requires less maintenance and adjustment than the older Strowger switch, it still has moving parts and metallic contacts which require occasional service attention. The cross-bar switch was therefore soon superseded by the reed relay switch. The reed relay comprises a pair of contacts made of a ferromagnetic material sealed in a glass tube as shown in Fig. 2.7. The glass tube is surrounded by a pair of coils so that, when current is passed through both coils simultaneously, a field is created which causes the reed contacts to move together. The residual magnetism in the contacts then causes them to remain closed until a demagnetising pulse is applied to one or other of the coils. A cross-point matrix can then be constructed out of reed relays by arranging a separate relay at each cross-point. Cross-point selection can be achieved by connecting one of the coil windings of each relay in series with its vertical neighbour and the other winding in series with its horizontal neighbour. The required cross-point is then selected by pulsing the appropriate vertical and horizontal circuit path simultaneously, thus closing the relay at the intersection point. It is then possible to select further cross-points by applying further pulses to the appropriate selection circuits. To release a selected cross-point, a reverse polarity pulse is applied to either the appropriate horizontal or the appropriate vertical selection line. In practice, each relay may contain up to four reed switches, enabling four-wire circuits to be switched where

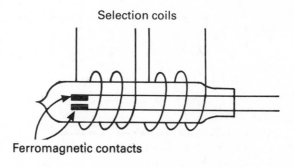

Fig. 2.7 – Reed relay switch contact.

necessary. The short pulses and low powers required to operate the contact points enable electronic circuits to be used to control the switching matrix as opposed to the traditional relay circuits required for step-by-step and cross-bar switches.

Having introduced the concept of the cross-point switch, we now consider the question of how to use them economically in a complete switching network. As with the Strowger two-motion switch, it is obviously impossible to fully interconnect every telephone subscriber using one enormous cross-point matrix. To simplify our discussion we shall refer to a switching matrix with M inputs and N outputs as an $M \times N$ switch. We shall represent an $M \times N$ switch matrix diagrammatically using the symbol shown in Fig. 2.8.

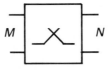

Fig. 2.8 – Symbol for $M \times N$ cross-point switch matrix.

As a first step, we can divide our M input circuits into x smaller groups of m inputs where

$$m = \frac{M}{x} \tag{2.1}$$

Similarly, we can divide our N output circuits into y smaller groups of n outputs each, where

$$n = \frac{N}{y} \tag{2.2}$$

We can then interconnect two stages of switching comprising x switching matrices of size $m \times y$ and y switching matrices size $x \times n$ as shown in Fig. 2.9. This obviously requires far fewer cross-points than a full $M \times N$ switch. However, it is only possible to set up a single circuit between any of the subscribers on a given input group and any of the subscribers on a single output group, although it is possible to connect the different subscribers in a single input group each to one subscriber in each of the output groups. Two cross-points have to be set up for each circuit.

The number of cross-points required.

$$C = x \times m \times y + y \times x \times m \tag{2.3}$$

Substituting equations (2.1) and (2.2) in (2.3) gives:

$$C = yM + xN \qquad (2.4)$$

The number of links between the first and second switching stages

$$L = x \times y . \qquad (2.5)$$

From (2.4), to minimise the number of cross-points required requires y and x to be as small as possible. However, if x and y are small, then L is small and there are few links available for switch interconnection. We therefore have to choose values of x and y which are as small as possible but give sufficient links to provide a reasonable grade of service to subscribers. Although it is possible to concentrate traffic by making $N < M$, or to expand the network availability by making $N > M$, most often switching stages, especially those at intermediate

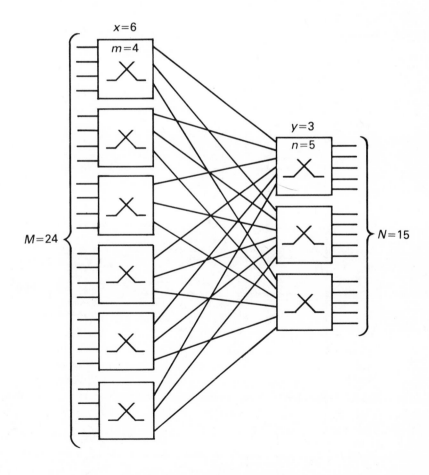

Fig. 2.9 – Two-stage switching matrix.

stages in the network, have as many outlets as inputs, that is, $N = M$. Under these circumstances it would be reasonable for the input and output groups to be similar size, that is $n = m$, and hence $x = y$. Since the same traffic is to be carried by the links as by the input and output circuits, it would not be un-reasonable to provide as many links as there are inputs and outputs to the switching system. Thus

$$M = N = L = xy = x^2 \; ,$$

Hence $\qquad\qquad\qquad x = \sqrt{M} \; .$

Since $m = M/x$, $m = \sqrt{M} = x$.

Thus a reasonable division of the input and output circuits is into $\sqrt{M}$

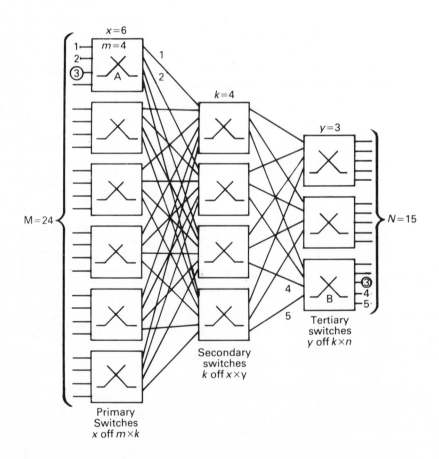

Fig. 2.10 – Three-stage switching network.

groups of $\sqrt{M}$ inputs/outputs each. This can only be properly achieved where M is a perfect square. In practice, integer values for m and x are chosen which are near to the optimum value of $\sqrt{M}$ and together give the necessary total number of inputs required.

The total number of cross-points then required, as given in equation (2.4) is

$$C = 2 M^{3/2}$$

The inability to connect more than one subscriber in a given input group to subscribers in the same output group arises because there is only one link between any given primary switch and any given secondary switch. Once this is in use, no further connection can be set up between two particular switches. This inability to connect otherwise free subscribers is known as blocking.

The probability of blocking can be greatly reduced by the use of three-stage switching. A generalised three-stage switching network is illustrated in Fig. 2.10. It is now possible to set up more than one circuit between a particular primary switch matrix and a particular tertiary switch matrix by using a separate route through each of the secondary switch matrices. It is still possible for blocking to occur, for instance, it would by impossible to connect a third input of primary switch A to a third output of tertiary switch B if the existing two inputs to primary A and the existing two outputs from tertiary B are all already connected to different secondary matrices. Under these conditions, all the secondary switches are already in use with regard to the relevant primary and tertiary switching matrices.

However, by providing sufficient secondary switching matrices, it is possible to construct a non-blocking three-stage switching network. Again, to simplify our analysis, we shall consider a network which has as many outlets as inputs and in which the input and output groups are of similar size. That is, we shall assume that $N = M$, $n = m$ and, hence, $x = y$. The worst situation, as far as blocking is concerned, is when all of the inputs but one on a particular primary switch A and all of the outlets but one on a particular tertiary switch B are each connected via a different secondary switch matrix as shown in Fig. 2.11. Under these circumstances, a further one secondary matrix is needed if it is required to connect the final input on primary switch A to the final output on tertiary switch B. Thus the requirement is for $2(m-1)+1 = 2m-1$ secondary switching matrices to be certain that blocking will not occur. The dimensions of the primary switches will be $m \times (2m-1)$, the secondary switches will be $x \times x$ and the tertiary switches $(2m-1) \times m$.

The total number of cross-points required is then:

$$C = 2 \times (m \times (2m-1)) + (2m-1)x^2$$

$$= \frac{4M^2}{x} - 2M + 2Mx - x^2$$

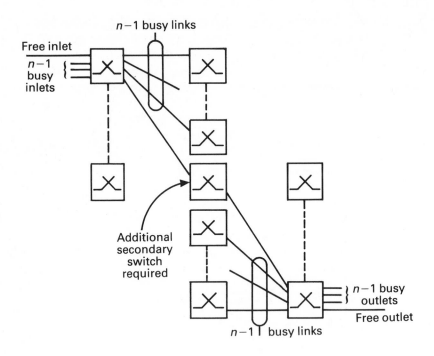

Fig. 2.11 – Non-blocking three-stage switching network.

We can then select x, the number of primary and tertiary switch blocks, to minimise C for a given M. For $M = 120$ the optimum value of $x = 15$. This gives a value of $C = 6975$. A fully provisioned single matrix would require 14,400 cross-points. The three-stage non-blocking network thus gives a saving of about half of the number of cross-points required. However, three switch selection operations instead of one are required for each through connection.

Except for certain military operations, the absolute non-blocking requirement is hardly ever necessary in practice. Networks with a small probability of blocking can be provided with much greater economy in the use of cross-points. However, the actual possibility of designing non-blocking networks which use fewer cross-points than a single stage switch is, in itself, an important principle.

ROUTING AND SWITCHING HIERARCHY

It is impossible for practical and economic reasons to fully interconnect all local telephone exchanges. Calls are therefore routed via a well-defined switching hierarchy. This hierarchy is illustrated in Fig. 2.12. Local exchanges are connected in groups to group switching centres (GSCs). Groups of GSCs are then

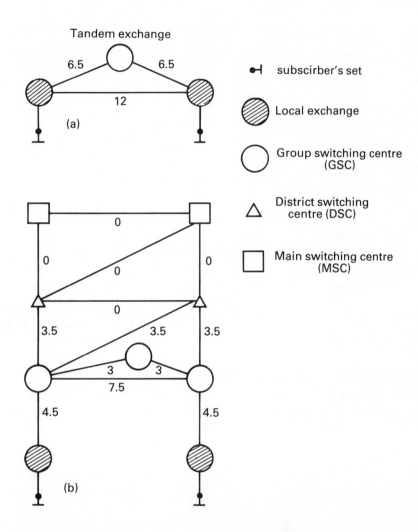

Fig. 2.12 — Routing and switching hierarchy. (a) Local traffic. (b) Longer-distance traffic. Notes: 1. Subscriber's lines and junctions are two-wire. 2. Trunk circuits are four-wire. 3. Switching at local exchanges and GSCs is two-wire. 4. Switching at DSCs and MSCs is four-wire.

connected to district switching centres (DSCs). The DSCs are then finally connected together in groups to main switching centres (MSCs). Only MSCs are fully interconnected. However, where considerable traffic exists between various stages of the hierarchy, appropriate circuits can be provided, as shown in Fig. 2.12. Interconnections between local exchanges and GSCs can be made via tandem exchanges provided specifically to aid effective use of interconnecting routes. The tandem exchanges have no local subscribers connected to them. An obvious case for interconnection at the local level is between exchanges located adjacent geographically but in separate operating areas such that they are served by different GSCs or DSCs or even different MSCs. This could mean that calls use only a mile or two of cable instead of many miles, perhaps even hundreds of miles, to separate MSCs and then between geographically widely separated MSCs. The numbers associated with the links in Fig. 2.12 are the maximum permissible losses in decibels. It will be noted that the permissible loss on the links at the top of the hierarchy is 0 dB. These are the longest links in the system and the transmission channels are therefore provided with amplifiers to compensate for the transmission loss. Because amplifiers are unidirectional by nature, transmission at this level must be by four-wire circuits. The DSCs and MSCs must therefore provide four-wire switching. The local exchanges and the GSCs require only two-wire switching as full-duplex transmission can be achieved with a single wire pair where amplifiers are not provided. Of course, in the upper layers of the routing hierarchy the transmission channels will almost certainly be provided by multiplex techniques. These are discussed in detail in a later chapter. The amplification will then be provided as an integral part of the multiplexing equipment. However, this does not effect the need for four-wire switching at the exchange, where each conversation still has to be dealt with individually.

DIGITAL SWITCHING SYSTEMS

Recent years have seen a rapid change to the use of digital techniques for the transmission of signals in the telephone network. This development is discussed later in Chapter 7. Another recent development in technology is the widespread availability of powerful computing facilities in the form of both large main-frame computers and microprocessors. Both of these developments have led to the possibility of using digital techniques for telephone switching. This offers enormous possibilities for enhanced services and facilities for the telephone user. Service flexibility is also possible in the form of stored program control of the switching network. This enables the network to be reconfigured and subscriber numbers to be re-allocated by simply keying in instructions by means of a keyboard associated with a visual display unit. Among the possibilities for enhanced services are 'follow-me transfer', 'camp-on busy' (call queueing), 'ring-back when free' and 'call forward to designated number'. This list is not exhaustive and customer demand will no doubt create a further range of services to be offered in the future. Digital switching is now being introduced into the UK telephone network under the generic title 'System X'. It is expected that the UK network will be entirely digital by the year AD 2000. We shall have to leave further discussion of digital switching, however, until we have looked at digital transmission in some detail. Digital switching techniques are therefore dealt with in more detail in the second half of Chapter 7 of this book.

3

Traffic theory

One of the problems of the telephone network engineer is to provide sufficient lines and switching equipment to give the customer a reasonable service, at the same time avoiding excessive cost through over-provision. To do this the engineer needs to know something of the nature of the traffic carried by the network. To do the job scientifically it is necessary to have some objective measures and some properly defined units.

The individual telephone user may use his telephone at any time, though there are certain times of the day when he is more likely to use it than at others. Taken over a large number of users, it is possible to obtain an estimate of the average pattern of calls throughout a typical day. Of course, this pattern varies from day to day. The behaviour of telephone users is different at weekends and on bank holidays from what it is on a normal working day. Events such as accidents or sudden weather changes can cause localised unpredictable changes in patterns of usage. A typical 24-hour call originations pattern is given in Fig. 3.1. The peak amplitudes are relative, units on the vertical scale will depend on the size of the sample of subscribers being considered. There is little use made of the network between 11.00 p.m. and 7.00 a.m., when most of the population is asleep. Business traffic causes a large peak around mid-morning, with a slightly smaller peak at mid-afternoon. The network is relatively lightly loaded during the normally accepted lunch-hour period. The peak of domestic calls occurs just after 6.00 p.m., when the evening cheap-rate call comes into effect. Clearly it is necessary to provide sufficient equipment to satisfy the

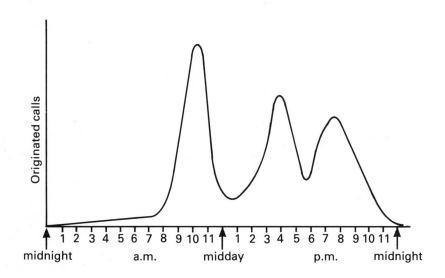

Fig. 3.1 – Typical 24-hour call originating pattern.

demand during the busiest period of the day. Provision is therefore made on the basis of the 'busy hour' statistics. From time to time the network will become overloaded, resulting in a degradation of service, owing to occasional unexpectedly high demands for telephone service. It would be uneconomical to provide equipment to satisfy the maximum possible demand. The provision is therefore made on the basis of the typical busy hour statistics and the occasional degradations are accepted as inevitable.

THE ERLANG

Before we can proceed further we need to define some measure of telephone traffic that we can use in subsequent calculations. In telephony, the word 'trunk' is used to describe a single connection between any one switching stage and another. A trunk may thus be a short connection between racks of switches or a connection between telephone exchanges in the network. In step-by-step switching there is usually a single switch mechanism at the terminating end of each trunk. It is therefore convenient, in calculations regarding the number of switching mechanisms required, to refer instead to the number of trunks required, since this avoids the necessity for multiple definitions of terms in our equations. Thus, in what follows, the term 'trunk' may be taken to include within its meaning 'switching unit', when such a meaning is appropriate to the calculations.

Now let us consider a single trunk. At any moment in time it will either be part of a single telephone circuit or it will be free. We shall define the traffic carried by a single trunk as the proportion of time that the trunk forms part of the telephone circuit, that is, it is carrying telephone traffic. Clearly this can be any value between 0 and 1. The unit of telephone traffic is the erlang, named after the Danish telephone engineer A. K. Erlang, whose paper on traffic theory, published in 1909, is now regarded as a classic. Thus, the maximum traffic that can be carried by a single trunk is 1 erlang. The traffic carried by a group of trunks is the sum of the traffic carried by each of the individual trunks. The maximum traffic that can be carried by n trunks is therefore n erlangs. The traffic carried by a group of trunks is thus the average number of calls in progress on that group of trunks at any particular instant of time. This, of course, does not help us to decide how many trunks we should provide in order to carry a given amount of traffic. It is only possible to carry n erlangs of traffic over n trunks if a new call commences as soon as an old call finishes, since each line must be fully occupied all of the time. We need, therefore to be able to compute the probability of there being x calls in progress at any given time and also the probability that all n trunks are already in use when another subscriber wishes to initiate a further call.

Before we can do this, we have to make some assumptions about the way in which calls arrive in the network. Two reasonable assumptions we shall make are:

(a) Individual calls occur on a purely random basis. That is, a subscriber is just as likely to make a call at one instant as at another.

(b) That the number of subscribers offering calls to the group of trunks under consideration is sufficiently large that the rate of call arrivals is not affected by the number of calls in progress in the group, that is, the system is in statistical equilibrium. Thus, the rate of call arrivals in any particular interval is independent of the arrival or non-arrival of calls outside this interval. Under these conditions the call arrivals are characterised by the Poisson distribution of statistics. According to the Poisson distribution, the probability of x calls arriving in an interval T is given by:

$$P_T(x) = \frac{(\alpha T)^x}{x!} e^{-\alpha T} , \qquad (3.1)$$

where α is the average calling rate (i.e. the average number of call arrivals in unit time).

When x is large, computation using the above equation can become tedious. Fortunately we can simplify the calculation by using Stirling's formula for $x!$ for large values of x. Stirling's formula states that:

$$x! = (\sqrt{2\pi x}) (x^x e^{-x}) .$$

If we substitute this in equation (3.1), we obtain the expression:

$$P_T(x) = \frac{1}{\sqrt{2\pi\alpha T}} e^{-(x - \alpha T)^2/2\alpha T} \tag{3.2}$$

which is, in fact, the Normal distribution, for which tables are widely available. Given that x is the number of calls initiated in unit time the traffic offered to a group of trunks is given by:

$$A = \alpha d , \tag{3.3}$$

where d is the average call duration.

Traffic represents the average number of calls in progress at any given time. Thus, even if the traffic offered to a group of trunks is less than the number of trunks in the group, the random nature of the origin of the calls means that occasionally an attempt will be made to set up a call when all the trunks are already occupied. In such a case, the attempt can be dealt with in one of two ways: either the call is put into a queue to await a free trunk or the attempt is aborted; the caller then making a fresh attempt to obtain a connection. The former is known as a delay system and the latter a loss system. Step-by-step switching is inherently a loss system.

ERLANG'S LOST-CALL FORMULA

The all-important question for a subscriber on a loss system is: what is the probability that I shall find all the trunks in use when I attempt to make a call? In other words, what is the probability that my call will be lost simply because there is not enough equipment available? This probability is given the somewhat misleading title of 'Grade of Service'.

To calculate the Grade of Service for a group of trunks we use Erlang's lost-call formula:

$$\text{Grade of service } B = \frac{\dfrac{A^N}{N!}}{\displaystyle\sum_{n=0}^{N} \dfrac{A^n}{n!}} ,$$

where A = traffic, in erlangs, and N = number of trunks in the group.

Erlang's formula is based on the assumptions made earlier regarding call arrivals in the network and that there is no restriction on the way in which calls can be allocated to particular trunks, that is, the trunks have 'full availability'.

Let us take an example to illustrate the use of the formula we have developed.

Example
A traffic load of two erlangs is offered to a full availability group of five trunks. The average call duration is three minutes.

(a) What is the average number of calls offered per hour?
(b) What is the probability that no calls are offered during a specified five-minute period?
(c) What is the Grade of Service (i.e. proportion of lost traffic)?
(d) If the trunks are always tested sequentially in the same order, how much traffic is carried by each trunk?

Solution
(a) Average number of calls per minute $x = A/d = 2/3$. Therefore average number of calls per hour $= 2/3 \times 60 = 40$.
(b) Using equation (3.1), probability of no calls arriving in a five-minute interval T is given by

$$P_s(0) = \frac{(2/3 \times 5)^0}{0!} \, e^{-(2/3 \times 5)}$$

$$= e^{-3.33}$$

$$= 0.0357$$

(c) Grade of Service (proportion of lost calls):

$$B = \frac{\dfrac{2^5}{5!}}{1 + \dfrac{2}{1!} + \dfrac{2^2}{2!} + \dfrac{2^3}{3!} + \dfrac{2^4}{4!} + \dfrac{2^5}{5!}}$$

$$= \frac{0.2083}{1 + 2 + 2 + 1.33 + 0067 + 0.2083}$$

$$= \frac{0.2083}{7.2083}$$

$$= 0.0289 \ .$$

Note that this represents an average of $0.0289 \times 40 = 1.156$ lost calls per hour.

(d) The proportion of traffic carried by the first trunk is equal to 1 minus the proportion of traffic lost to the first trunk. We can calculate this by using Erlang's formula for a single trunk.

Thus proportion of traffic carried by first trunk

$$C_1 = 1 - B_1$$

$$= 1 - \frac{\dfrac{2}{1}}{1 + \dfrac{2}{1}}$$

$$= 1 - 0.67 = 0.33 \ .$$

Therefore traffic carried by first trunk:

$$A_1 = C_1 A$$

$$= 0.33 \times 2$$

$$= 0.67 \text{ E} \ .$$

The remaining traffic, $2 - 0.67 = 1.33$ E, will be offered to the second trunk.

We can now carry out similar calculations for the traffic carried by the re-maining trunks, although we must use Erlang's formula for 1, 2, 3 etc. trunks successively, each time taking the traffic offered to the group as 2 E. It is not valid to consider each trunk as a separate single trunk being offered the traffic lost to the preceding trunk, since the traffic passed on in this way may not necessarily have the proper conditions of randomness to validate the use of Erlang's formula.

Thus the proportion of traffic lost to each subsequent trunk can be shown to be:

$B_2 = 0.4$
$B_3 = 0.2105$
$B_4 = 0.0952$
$B_5 = 0.0289$ – (this is the Grade of Service of the whole group).

Assuming the 'lost' traffic is passed on sequentially to the network, the traffic carried by each trunk will be as follows.

$$
\begin{aligned}
A_2 &= A\,(B_1 - B_2) &&= 0.54 \text{ E}\\
A_3 &= A\,(B_2 - B_3) &&= 0.38 \text{ E}\\
A_4 &= A\,(B_3 - B_4) &&= 0.23 \text{ E}\\
A_5 &= A\,(B_4 - B_5) &&= 0.13 \text{ E}\\
\text{The total traffic lost} && &= 0.05 \text{ E}
\end{aligned}
$$

Note that each succeeding trunk carries less traffic than its predecessor. The loss of a trunk through fault conditions means that the group is effectively reduced in size by one trunk. The practical effect is to lose the last trunk in the sequence from the standpoint of extra traffic lost to the network (i.e. degradation in the grade of service). The temporary loss of a single trunk may not therefore be as disastrous as would appear at first glance.

QUEUEING SYSTEMS

The traffic concepts we have considered so far are based on a loss system. That is, any calls unable to find a route through the network are aborted and lost to the system. Of course the caller may make a further attempt to establish the call, but the statistics of call origination are such that it is justifiable to regard this as an entirely new attempt, statistically independent of the previous abortive attempt. The step-by-step switching concept, which until the recent advent of digital switching systems has enjoyed almost universal use, is a strictly loss system. There is no mechanism by which calls can be queued until equipment is available. However, the introduction of digital switching has now led to the possibility of calls being held in waiting in queues until equipment or circuits become available to establish the call connection. We shall therefore take a brief look at the effect of queueing on traffic considerations.

The effect of queueing is that a call will be delayed rather than lost. However, with a lost call system there is always a faint possibility, however small, that a free trunk will be found, no matter how few trunks exist and however great the busy-hour traffic. With queueing the situation is different. If the traffic offered is greater than the number of trunks, or 'servers' as they are usually referred to in delay systems, then the queue will forever be increasing in length and the probability that a call will be delayed will become unity, at the same time the length of the delay will be always increasing. Because there is always a possibility there are others waiting in the queue, the probability of delay is always greater than the probability of a lost call in the equivalently dimensioned system. It is possible to derive a formula, although the derivation will not be given here, which relates the probability of delay D to the probability of a lost call B in the equivalently dimensioned system. The formula is:

$$D = \frac{NB}{N - A\,(1 + B)}$$

where B = Grade of Service, from equation (3.4), N = number of servers (trunks), and A = traffic offered in erlangs

This equation is known as Erlang's delay formula. Its validity is based on the same assumptions as Erlang's lost-call formula, plus:

(a) the queue can accept an infinite number of waiting calls;

(b) the queue is served on a first-in-first-out basis, i.e. calls are dealt with in the order of arrival.

The first of these assumptions is practically unrealisable. However, providing N is somewhat greater than A, then many real situations will have sufficient queue facilities for overflow to be rare enough to make the analysis valid for all practical purposes. If A is greater than N, then the queue must always be increasing in length and the system is therefore no longer stable.

Other parameters besides probability of delay are of interest. Among these are those relating to queue length and waiting time. These problems, however, are not readily amenable to analytical solution and computer simulation is therefore normally used to determine the performance of any proposed system. The results of such simulations are generally system-dependent and it is therefore impossible to give meaningful data in this elementary discussion of the general principles of traffic theory. However, sufficient has been said to indicate the sort of trade-off that exists between the frustration of having to occasionally make repeated attempts to establish a call on a lost call system and that of sometimes having to queue for a significant period of time on a delay system involving call queuing.

4

Transmission of telephone signals

So far we have only considered the problem of telephony in relation to the need to switch the signals so that subscribers can be interconnected with one another at will. We shall now leave the problems of switching to consider exactly how the signals are conveyed between the various interconnecting switching stages.

In a very basic telephone network, all that is required for each connection is a single switched circuit consisting of a simple pair of wires connecting end to end. In practice, however, it is only economical to provide separate wire pairs at the local ends of telephone circuits. For the long-distance interconnections between telephone exchanges, it is usually more economical for several telephone circuits to share a single wire connection using multiplexing techniques. In this chapter we shall consider how economical use can be made of the available transmission facilities.

TRANSMISSION LINE THEORY

To help us to understand something of the nature of telephone lines, we shall begin with a brief study of transmission line theory. A transmission line consists of a pair of conductors. Since the conductors are made from a physical material, such as copper or aluminium, the conductors themselves will have resistance and inductance. Between the conductors there will be leakage conductance and capacitance. We can consider a line of length l to be made up of many elemental

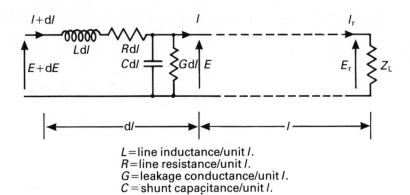

L=line inductance/unit l.
R=line resistance/unit l.
G=leakage conductance/unit l.
C=shunt capaçitance/unit l.

Fig. 4.1 – Equivalent circuit of transmission line.

sections of length dl as shown in Fig. 4.1 From the diagram, we see that

$$dE = I(R + j\omega L)dl \tag{4.1}$$

and

$$dI = E(G + j\omega C)dl \tag{4.2}$$

where ω is the radian frequency of the signal voltage E.
  If we let $R + j\omega L = Z$ and $G + j\omega C = Y$, then from (4.1) and (4.2):

$$\frac{dE}{dl} = (R + j\omega L)I = ZI \tag{4.3}$$

and

$$\frac{dI}{dl} = (G + j\omega C)E = YE \tag{4.4}$$

Differentiating (4.3) and substituting (4.4), we get:

$$\frac{d^2E}{dl^2} = Z\frac{dI}{dl} = ZYE \tag{4.5}$$

Similarly

$$\frac{d^2I}{dl^2} = ZYI . \tag{4.6}$$

Solving these equations, we get:

$$E = E_1 e^{\sqrt{ZY}\,l} + E_2 e^{-\sqrt{ZY}\,l} \tag{4.7}$$

and

$$I = I_1 e^{\sqrt{ZY}\,l} + I_2 e^{-\sqrt{ZY}\,l} . \tag{4.8}$$

Substituting (4.8) in (4.4):

$$YE = ZY I_1 e^{\sqrt{ZY}\, l} - ZY I_2 e^{-\sqrt{ZY}\, l}$$

therefore

$$E = \frac{Z}{Y} I_1 e^{\sqrt{ZY}\, l} - \frac{Z}{Y} I_2 e^{-\sqrt{ZY}\, l}.$$

Comparing with (4.7), we get:

$$I_1 = \frac{E_1}{\sqrt{Z/Y}} \quad \text{and} \quad I_2 = \frac{-E_2}{\sqrt{Z/Y}},$$

therefore from (4.8), we get:

$$I = \frac{E_1}{\sqrt{Z/Y}} e^{\sqrt{ZY}\, l} - \frac{E_2}{\sqrt{Z/Y}} e^{-\sqrt{Z/Y}\, l} \tag{4.9}$$

From (4.8) and (4.9) we can see that E_1 is an 'incident' wave and E_2 a 'reflected' wave.

The term $\sqrt{ZY}$ is usually referred to as the propagation constant γ. Thus, in terms of distributed parameters of the line

$$\gamma = \sqrt{ZY} = \sqrt{\{(R+j\omega L)(G+j\omega C)\}} \tag{4.10}$$

Note that γ is generally complex and may be separated into real and imaginary parts such that $\gamma = \alpha + j\beta$.

If the line is lossless, that is there is no series resistance or leakage conductance, the γ is purely imaginary. The term α thus represents the losses on the line and is therefore known as the attenuation constant. Its units will be in nepers/unit length, 1 neper being equal to 8.686 dB. Likewise, the term β represents the phase displacement of the signal along the line and is known as the phase constant. Its units will be in radians per unit length. The wavelength of the line signal is the distance along the line between adjacent points of identical phase of the incident wave, that is, a phase displacement of 2π radians. Thus the wavelength $\lambda = (2\pi)/\beta$ and we can define a 'velocity of phase propagation' $v_p = \lambda f$. For a lossless line $\beta = \omega\sqrt{LC}$ and $v_p = 1/\sqrt{LC}$. For a lossy line, both α and v_p are generally frequency dependent.

Now we shall evaluate the boundary conditions for equations (4.8) and (4.9) where $1 \to 0$, that is, at the load Z_L where $E \to E_r$ and $I \to I_r = E_r/Z_L$. This gives us:

$$E_r = E_1 + E_2 \tag{4.11}$$

and

$$I_r = \frac{E_1}{\sqrt{Z/Y}} + \frac{E_2}{\sqrt{Z/Y}}. \tag{4.12}$$

The term $\sqrt{Z/Y}$ is known as the 'characteristic impedance' Z_0 of the line. It represents the impedance seen at the input of a line of infinite length. Note that it has a purely resistive value only if the line is lossless, that is $R = 0$ and $G = \infty$. For a lossless line the characteristic impedance will be given by the resistance $R_0 = \sqrt{L/C}$.

In a lossy line the characteristic impedance is frequency dependent. However, as the frequency is increased $Z_0 \to \sqrt{L/C}$. This is the value of Z_0 normally quoted by cable manufacturers.

From (4.12):

$$I_r = \frac{E_1}{Z_0} - \frac{E_2}{Z_0} ,$$

therefore

$$Z_0 I_r = E_1 - E_2 . \tag{4.13}$$

Adding (4.10) and (4.12), we get:

$$Z_0 I_r + E_r = 2E_1$$

therefore

$$E_1 = \frac{E_r + Z_0 I_r}{2} = \frac{E_r}{2}\left(1 + \frac{Z_0}{Z_L}\right) .$$

Similarly, subtracting (4.13) from (4.11), we get:

$$E_2 = \frac{E_r - Z_0 I_r}{2} = \frac{E_r}{2}\left(1 - \frac{Z_0}{Z_L}\right) .$$

Note that if we terminate the line with its characteristic impedance, that is, $Z_L = Z_0$, there will be no reflected wave E_2. On the other hand, if we short-circuit the line, that is, we make $Z_L = 0$, $E_1 = -E_2$ and $E_r = 0$. Similarly, if we leave the line open-circuit, that is, we make $Z_L = \infty$, then $E_1 = E_2$ and $E_r = 2E_1$. Thus there will be reflected waves on lines that are not terminated with their characteristic impedance. We can define a reflection coefficient ρ, where

$$\rho = \frac{E_2}{E_1} = \frac{Z_L/Z_0 - 1}{Z_L/Z_0 + 1} = \frac{Z_L - Z_0}{Z_L + Z_0} .$$

Because the incident wave and the reflected wave are travelling in opposite directions along the line, they will add or subtract according to their relative phases at a given point along the line. There will therefore be nodes and antinodes of voltage and current, forming a standing wave pattern along the line. The distribution of voltage and current along a lossless line for various terminations is given in Fig. 4.2. We can therefore define a 'standing wave ratio' (SWR)

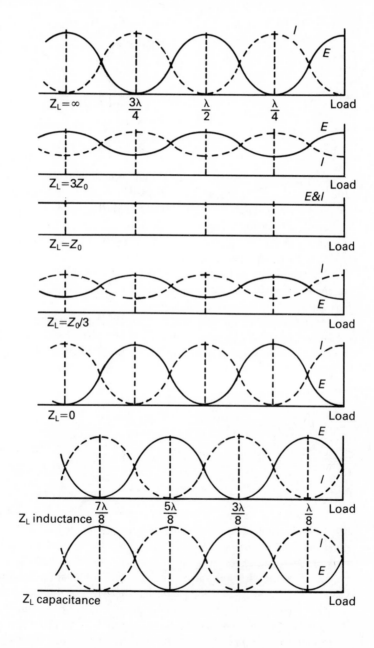

Fig. 4.2 – Standing wave patterns for various values of load Z_0.

in terms of the ratio of the maximum and minimum value of the standing wave voltage along the line. Thus the voltage standing wave ratio σ is given by:

$$\sigma = \frac{E_{max}}{E_{min}} = \frac{E_1 + E_2}{E_1 - E_2}$$

$$= \frac{1 + |\rho|}{1 - |\rho|} .$$

Hence

$$|\rho| = \frac{\sigma - 1}{\sigma + 1} .$$

It is obviously desirable that each line should be terminated in its characteristic impedance. The characteristic impedance of typical telephone cable wire pairs is between about 600 and 1000 Ω. Telephone speech signals are normally limited to a frequency range of about 300 to 3400 Hz and at these frequencies the wavelength of the signal on the line is many times greater than the length of a typical cable run. Thus, even with a significant load mismatch, there will not be very significant variations in the amplitude of signals along the line due to standing waves. The matching is therefore less critical than it would be for high-frequency or long-distance applications.

A typical value for the attenuation constant for telephone wire pairs is between 2 and 3 dB per mile. From an examination of equation (4.10), it can be seen that it is possible to minimise α with respect to the distributed inductance L. It can be shown that the value of L for minimum α is:

$$L = \frac{RC}{G} ,$$

giving a value of attenuation constant:

$$\alpha = \sqrt{RG} .$$

In practice L is usually lower than the optimum value and it is possible either by distributed loading or by the use of lumped inductance at intervals along the line, to artificially increase L towards its optimum value. On cable pairs of lengths in excess of about 3½ miles, a common practice is to fit 88 mH coils at intervals of 2000 yards. The resultant attenuation is about 1 dB/mile, although the use of fixed loading coils has the effect of producing a sharp cut-off at higher frequencies. For the example given, this cut-off occurs at about 3500 Hz. The effect of loading on attenuation is illustrated in Fig. 4.3.

A bonus in optimising L in this way is that, under these conditions, the velocity of phase propagation:

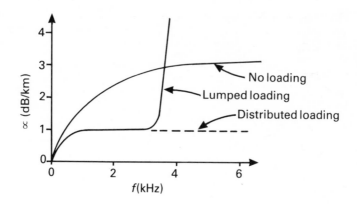

Fig. 4.3 – Attenuation characteristics for typical wire-pair lines.

$$v_p = \frac{1}{\sqrt{LC}}$$

as for the lossless line and is therefore independent of frequency. This means there is no distortion arising from differential propagation of the various frequency components of the speech signal.

Individual wire pairs are generally used for local line transmission between subscriber and local exchange. For short-distance inter-exchange working, up to about 20 miles, separate wire pairs in multipair loaded cables are used for each conversation. For longer distances several conversations are multiplexed onto each cable pair. Until the comparatively recent advent of pulse-code-modulation, this was done using frequency-division-multiplex (FDM) techniques.

FREQUENCY-DIVISION-MULTIPLEX

The bandwidth of a telephone speech signal is rather less than 4 kHz whereas the available bandwidth on unloaded cable pairs is well above 100 kHz. It is therefore possible, using modulation techniques, to divide up the cable bandwidth so that a number of telephone speech paths can be carried simultaneously along a single cable pair. The normal arrangement consists of 24 telephone channels per cable pair, the modulation into 24 channels being carried out in two stages. In the first stage 12 channels are multiplexed together to form what is commonly known as a basic group. The basic group arrangement is illustrated in Fig. 4.4. Each of the 12 telephone signals are single-side-band amplitude modulated onto carriers spaced at 4 kHz intervals from 64 kHz to 108 kHz. The lower side-band (LSB) is used in each case. The 12 base-band signals are therefore translated into the frequency band from 60 kHz to 108 kHz as

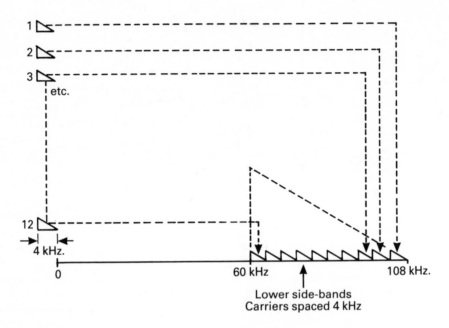

Fig. 4.4 – Basic group arrangement.

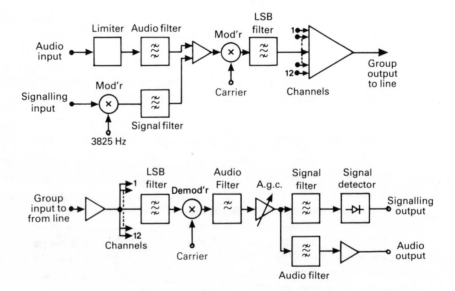

Fig. 4.5 – Channel translating equipment.

shown. The block diagram for the channel translating equipment is given in Fig. 4.5. To form a 24-channel system, two basic groups are taken together. One basic group (B) is transmitted directly as it stands. The other basic group (A) is amplitude modulated onto a carrier ar 120 kHz and the lower side-band is taken so as to occupy the frequency range from 12 kHz to 60 kHz as shown in Fig. 4.6. By using two stages of modulation for basic group A, it is possible to reduce the physical size of the components required for the LSB filters of Fig. 4.5, since the lowest cut-off frequency required is at 64 kHz rather than 12 kHz if the whole block of 24 channels were assembled together in one stage of modulation.

One of the difficulties with single-side-band modulation is that it is impossible to transmit a d.c. component in the base-band signal. On cable-pair circuits the channel signalling is carried out by means of a d.c. signal derived from the exchange battery. To allow signalling information to be conveyed through the FDM channel, 'outband' signalling is used. Each channel is allocated a 4 kHz bandwidth slot, whereas the basic telephone speech signal only occupies a bandwidth from about 3000 to 3400 Hz. It is therefore possible to include a signalling tone in the channel slot above the highest frequency in the telephone speech signal. A frequency of 3825 Hz is chosen for this signal. This tone can be separated from the audio speech signal by means of filters as shown in Fig. 4.5. The 3825 Hz tone is switched on wherever the associated telephone circuit has a d.c. signalling level present.

SUPER-GROUPS

Occasionally it is possible to select a cable pair that has an available bandwidth considerably in excess of 100 kHz. Alternatively, on long-distance routes it can be economical to lay coaxial cables in place of multipair cables having a bandwidth capability of several megahertz. Under these circumstances it is

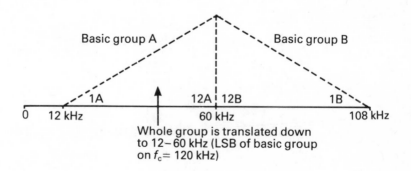

Fig. 4.6 – 24-channel FDM allocation.

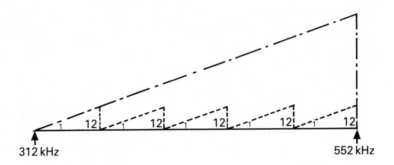

Fig. 4.7 – Basic super-group (60 channels).

possible to group together much larger numbers of channels onto a single transmission path. To obtain a 60-channel 'super-group', five basic groups are single-
side-band amplitude modulated onto carriers spaced at 48 kHz intervals from
420 kHz to 612 kHz so that the lower side-bands fill the frequency band from
312 kHz to 552 kHz as shown in Fig. 4.7. In fact, the 60-channel super-group
is rarely used on its own as it stands. Two super-groups are taken together to give
120 channels as shown in Fig. 4.8. One super-group (2) is transmitted directly
as it stands. The other, (1), is single-side-band amplitude modulated onto a
carrier at 612 kHz so that the lower side-band fills the frequency range from 60
kHz to 300 kHz as shown. A 'guard interval' is left between 300 kHz and 312
kHz to ease the design of filters in the modulation apparatus. If only 60 channels
are required, it is more common to modulate the basic super-group down to the
60 kHz to 300 kHz frequency range than to leave the basic super-group in its
initial frequency band.

Occasionally five super-groups are assembled together in a similar way to
form a 300-channel 'master' group. Finally, 16 super-groups are sometimes

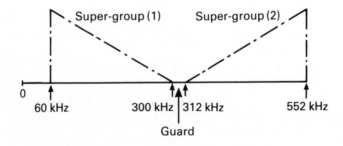

Fig. 4.8 – 120-channel FDM allocation.

assembled together to give 960 channels occupying a bandwidth of approximately 4 MHz.

NOISE

The frequency-dependent characteristics of the transmission line are not the only thing to impair the transmission of signals along the line. The transmission is also impaired by the presence of unwanted extraneous signals on the line. These may occur in a number of ways. Interference signals can be induced in the line, either from external sources such as lightning flashes or vehicle ignition, or from crosstalk between adjacent channels in the form of interfering speech signals or switching impulses. Random noise also occurs naturally in components in the speech path circuit, especially from repeater amplifiers. We shall look at these noise sources in turn, commencing with the random 'background' noise produced by the circuit components. This is the noise heard as a faint hiss in the background of any telephone conversation. Although it can be troublesome if it reaches significant proportions, a little background noise is, in fact, desirable, as it gives the user some confidence that his equipment is not dead.

Any electronic component produces 'noise'. The term noise is used here to include any unwanted signal, whether audible or not. This noise can extend into the radio frequency spectrum. Let us consider the most basic of all components, the resistor. Noise will be produced in the resistor by the random motion of electrons within the resistive material. Since the mobility of the electrons is a function of temperature, we would expect the noise power to increase with temperature and to become zero only at absolute zero temperature. Such noise is referred to as thermal or Johnson noise. The mean square thermal noise voltage at the terminal of an open-circuit resistor of value R_s is:

$$V_n^2 = 4kTBR_s$$

where T is the temperature of the resistor, in Kelvins, B is the noise bandwidth,

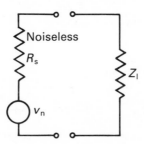

Fig. 4.9 – Equivalent circuit of thermal noise in R_s.

and k is Boltzmann's constant $= 1.38 \times 10^{-23}$ J/K.

The power spectral density of thermal noise is constant over a very wide frequency range and is therefore often referred to as white noise. The noise bandwidth is therefore determined by the bandwidth of the circuit over which the noise signal will be transmitted. An equivalent circuit for the resistor incorporating the noise voltage source is given in Fig. 4.9. If the source is connected to a load Z_L, then the noise power transferred to the load is:

$$P_n = \frac{V_n^2}{(R_s + Z_L)^2} \times Z_L \, .$$

Maximum power transfer occurs when $Z_L = R_s$, then

$$P_n = \frac{V_n^2}{(2R_s)^2} \times R_s = \frac{4kTB \, R_s^2}{4R_s^2} = kTB \, .$$

This value is known as the available power. If $Z_L \neq R_s$, then the power transferred will be less than the available noise power.

Any white noise source can be specified in terms of an effective noise temperature, even if the fundamental origin of the noise is other than thermal. We will therefore define an effective noise temperature T_s where

$$T_s = \frac{\text{noise power delivered by source in bandwidth } B}{kB} \, .$$

Similarly, we can define an effective noise temperature of an amplifier T_A where:

$$T_A = \frac{\text{noise power output due to generation within the amplifier}}{GkB_n} \, .$$

where G is the amplifier power gain and B_n is the noise bandwidth of the amplifier.

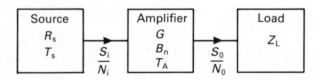

Fig. 4.10 – Effective noise temperature model.

To illustrate the use of these effective noise temperatures, we will consider the amplifier of gain G and noise bandwidth B_n illustrated in Fig. 4.10.

The input noise power to the amplifier from the source $N_i = kT_sB_n$. If the signal input power to the amplifier is S_i, then input signal-to-noise ratio

$$\frac{S_i}{N_i} = \frac{S_i}{kT_sB_n} \quad .$$

The output noise power delivered to load,

$$N_0 = GN_i + GkT_AB_n = GkB_n\,(T_s + T_A) \quad . \tag{4.14}$$

Thus the noise generated within the amplifier has been converted into a corresponding increase in the effective temperature of the source. Thus the output signal-to-noise ratio.

$$\frac{S_0}{N_0} = \frac{GS_i}{GkB_n(T_s + T_a)} = \frac{T_s}{T_s + T_a}\left(\frac{S_i}{N_i}\right) \quad . \tag{4.15}$$

Although we have taken an amplifier as our example, the above equations are equally applicable to any two-port network. In the case of an attenuator, or a lossy transmission line, G will be less than unity. If a passive two-port network, such as a transmission line, reduces the signal power transmitted by a factor L, that is $L = 1/G$, then its noise temperature is:

$$T_A = (L-1)T_0 \quad ,$$

where T_0 is the ambient temperature of the network.

It is usual to take a standard value for the ambient temperature of 290 K, this being roughly equivalent to a standard room temperature of $17°C$. Thus, for a transmission line having a loss of 2 dB, $L = 1.585$ and hence:

$$T_A = (1.585 - 1) \times 290 = 169.6\,K \quad .$$

Noise temperatures are normally only used to describe low noise devices for which $T_A \ll T_s$. In other cases the noise factor F is used instead. We shall define F as:

$$F = \frac{\text{available noise power at two-port output}}{\text{available noise power output if two-port were noiseless}} \quad ,$$

the source being at the standard ambient temperature $T_0 = 290$ K. Then

$$F = \frac{N_0}{}$$

the nomenclature being the same as for the equations above.

Since we have taken $T_s = T_0$, from (4.14),

$$N_0 = Gk(T_0 + T_A)B_n \ .$$

Thus

$$F = \frac{T_0 + T_A}{T_0} \ \text{and} \ T_A = (F-1)T_0 \ ,$$

and hence, from (4.15):

$$\frac{S_0}{N_0} = \frac{1}{F}\left(\frac{S_i}{N_i}\right) \ .$$

Let us now consider a number N of matched two-ports connected in cascade. Let the individual power gains, effective noise temperatures and noise factors for each two-port be:

$$G_1, G_2 \ldots G_N; \quad T_1, T_2 \cdots T_N; \quad F_1, F_2 \cdots F_N \ .$$

The overall power gain of the system will be $G_s = G_1 G_2 \ldots G_N$.

If the system noise band-width is B_n and a source having an effective noise temperature T_s is connected to the input, then the noise output from the system,

$$N_0. = G_s k T_s B_n + G_s k T_1 B_n + (G_2 G_3 \ldots G_N) k T_2 B_n + \ldots + G_N k T_N B_n$$

$$= G_s k T_s B_n + G_s k B_n (T_1 + \frac{T_2}{G_1} + \frac{T_3}{G_1 G_2} + \ldots + \frac{T_N}{G_1 G_2 \ldots G_{N-1}}$$

$$= G_s(T_s + T_e) \ ,$$

where

$$T_e = T_1 + \frac{T_2}{G_1} + \frac{T_3}{G_1 G_2} + \ldots + \frac{T_N}{G_1 G_2 \ldots G_{N-1}}$$

is the effective noise temperature of the whole cascade.

We can, of course, alternatively express the noise output in terms of the individual two-port noise factors. In this case:

$$N_0 = G_s k T_s B_n + G_s k B_n (F_1 - 1) T_0 + (G_2 G_3 \ldots G_N) k B_n (F_2 - 1) T_0 + \ldots$$

$$= G_s k T_s B_n + G_s k T_0 B_n \left(F_1 - 1 + \frac{F_2 - 1}{G_1} + \frac{F_3 - 1}{G_1 G_2} + \ldots \right) .$$

If the source is at $T_s = T_0$, then:

$$N_0 = G_s k B_n T_0 \left(F_1 + \frac{F_2 - 1}{G_1} + \frac{F_3 - 1}{G_1 G_2} + \ldots + \frac{F_{N-1}}{G_1 G_2 \ldots G_N} \right)$$

whence the overall noise factor of the system:

$$F_e = F_1 + \frac{F_2 - 1}{G_1} + \frac{F_3 - 1}{G_1 G_2} + \ldots + \frac{F_{N-1}}{G_1 G_2 \ldots G_N})$$

and the overall system noise output is:

$$N_0 = G_s k B_n F_e T_0 .$$

Let us consider the problem of transmitting an analogue signal, such as a speech signal, over a path with a finite transmission loss and subject to noise impairment. As the length of the transmission path is increased, the receive signal-to-noise ratio, for a given transmitted signal level, decreases. Alternatively the signal may be completely destroyed by the additive noise.

One way to overcome this problem would be to increase the transmitted signal level to such an extent that even the attenuated received signal is large enough to overcome the noise. There are, however, practical difficulties with such a suggestion. For example, for a 3000-mile cable with an attenuation of 1 dB per mile (a typical figure for telephone cable) an input signal of 10^{147} volts is required for a received signal level of 1 mV. An amplifier at the receiver is no help as this would increase both signal and noise together. Amplifiers at points along the line can improve the situation, as illustrated in Fig. 4.11. An ultimate point is set, however, which is equivalent to the lossless line. Above this, it is impossible to improve the receive signal-to-noise ratio. In fact, in practice, even this limit cannot be achieved as noise will be added to the system by the amplifiers themselves. In the next chapter we shall see from Shannon's theorem that it should be possible, by increasing the bandwidth of our signal, to convey the same information with a lower signal-to-noise ratio. One way in which this can be done is to sample our original signal and represent the sample amplitudes digitally, usually in binary digits. This will indeed increase the bandwidth of our signal, as we shall see later, but now we only have to be able to recognise the presence or polarity of pulses in a pulse train. This we can do at levels of noise which would certainly have completely masked our original

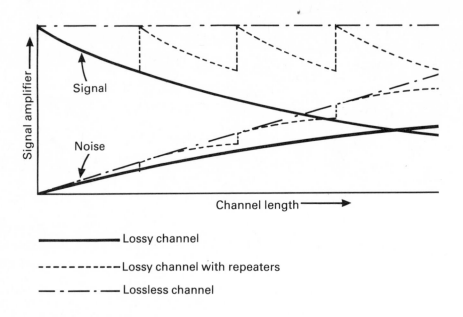

Fig. 4.11 – Channel noise impairment.

signal. What is more, instead of intermediate amplifiers, which increase the noise as well as the signal, we can install regenerators along the line which regenerate the pulse-train entirely free from noise. We thus have a technique which will enable us to transmit signals over channels which would otherwise be useless because of the noise impairment to the signal. Such a system, proposed by A. H. Reeves in 1939, is known as pulse-code-modulation, or simply as PCM. However, before we can make a more detailed study of PCM, we first need to take a closer look at information theory and the principles of digital transmission.

5

Information theory and coding

For almost seventy years following the invention of the telephone, telecommunications was almost synonymous with telephony. However, with the emergence in the 1950s of computers capable of large-scale data processing, a need for data communications began to emerge. Now, in the 1980s information technology is increasingly becoming an important, if not quite yet an essential, part of the modern way of life. As it does so, it creates an ever-increasing demand on the communications engineer for more complex and more efficient information transmission systems.

To be able to design such systems, the engineer needs first to reach an understanding of the nature of information and the ways in which it can be represented. We need, therefore, some measure of information so that we can reach some objective conclusions regarding the efficiency of its transmission. For this, we shall need to turn to C. E. Shannon, whose work on information theory dates from around 1948. It is no exaggeration to say that Shannon's work revolutionised thinking on information transmission. Before Shannon, it was generally thought to be impossible to transmit information and receive it with absolute certainty over a noisy communication channel. Shannon's work shows that it is in fact quite possible to transmit information at a finite rate over a noisy communication channel with an arbitrarily small probability of error.

To the general public, a measure of information would need to include some value judgement regarding the significance of an event. The winning premium bond number will obviously be of much greater interest to a holder of several bonds than to the person who has no bonds and has no intention of purchasing any. To the communications engineer, this value judgement is of no significance. To him, the winning number is just one message that can be selected from a set of many possible messages. Thus Shannon's theory is concerned with the statistical properties of symbols or messages that can be selected from a suitably defined set, known as an alphabet or an ensemble. The problem then is to represent these messages in such a way that they can be transmitted as efficiently as possible with the minimum likelihood of error.

INFORMATION SOURCES

There are basically two types of information source, discrete and continuous. In a discrete source, messages or symbols are selected from a finite set $x_1, x_2, \ldots, x_n$ according to some probability rule. Telex is a simple example of a discrete source, where letters of the alphabet are selected to form a textual message.
The probabilities of occurrence of the individual letters are determined by the properties of the language used to convey the text. In a continuous source the selections are made from a set of values which is continuous within a given range. Thus the value of voltage as indicated by a meter pointer is an example of a continuous information source.

There are two other source classifications we need to consider before we move on to consider a measure of information. We shall normally restrict our study to what are known mathematically as 'Ergodic sources'. An ergodic source is one in which every sequence of symbols produced has the same statistical properties; that is the statistics are 'stationary'. This means that all strings of symbols produced by the source, provided they are of sufficient length to be statistically significant, will contain similar proportions of individual symbols and symbol combinations. Thus any particular language is likely to contain a similar proportion of each letter of the alphabet and of letter combinations such as th, st, ou; although the particular combinations may change if the language is changed. For example, whereas the letter combinations pf and zw occur quite frequently in German, they hardly ever occur as combinations in English.

The other classification of sources is into memory-less sources and those with memory. A memory-less source, or sometimes called a source with zero memory, is one in which each selection is completely independent of all previous selections. If the previous selections influence the present selection the source is said to have memory. A source with memory is often referred to as a Markov source. If the selection is influenced by the previous m selections, then the source is referred to as an mth order Markov source. Language is obviously a source with memory, for it is frequently possible to predict the next letter in a sequence of letters from a word in the language. For example, in English u will nearly always follow q and the longer a sequence of consonants, the more likely it is that a vowel will follow.

A MEASURE OF INFORMATION

We require a measure of information that is related to the probability of the occurrence of an event. When two events occur, the information related to the two events should be additive. The joint probability of two events is the product of the two individual probabilities. A logarithmic function of probability is therefore a desirable unit of measure of information. Since the probability of occurrence of an event must always be less than unity, the logarithm of the probability would be negative. It is desirable that our measure of information should be in positive units. A reasonable measure is therefore:

$$I(x_i) = \log_a \frac{1}{P(x_i)}$$

where $I(x_i)$ is the information associated with the occurrence of event x_i and $P(x_i)$ is the probability of the occurrence of the event.

The base of the logarithm will determine the units of information. The most widely used unit is that of 'bits', a shortened form of binary digits, where $a = 2$. Alternatives are nats (or nits), a shortened form of natural units, when the logarithms are taken to the base e, and hartleys, after R. V. L. Hartley, when the logarithms are taken to base 10. Thus

$$1 \text{ hartley} = 3.322 \text{ bits}$$

and

$$1 \text{ nat} = 1.443 \text{ bits.}$$

Note that the information associated with an event increases as its likelihood decreases. Thus we gain a lot of information when we learn that a rare event has happened, whereas we gain little information when we learn that what has happened was what we had rather expected to happen. This accords with how we naturally regard information in ordinary day-to-day experience. It thus seems a very reasonable basis on which to proceed as we consider the subject of information theory in more detail.

INFORMATION FROM A ZERO-MEMORY SOURCE

If messages are selected by a zero-memory information source from a message set $x_1, x_2, \ldots, x_n$, where the probabilities of selecting the messages are $P(x_1)$, $P(x_2), \ldots, P(x_n)$, respectively, then the information generated each time the message x_i is selected is

$$\log_2 \frac{1}{P(x_i)} \text{ bits.}$$

The message x_i however, will, on average, be selected a proportion $P(x_i)$ of the total number of selections, providing the total number of selections is large enough for statistical significance. Thus the average amount of information per message selection, H, is given by:

$$H = P(x_1)\log_2 \frac{1}{P(x_1)} + P(x_2)\log_2 \frac{1}{P(x_2)} + \ldots + P(x_n)\log_2 \frac{1}{P(x_n)}$$

$$= \sum_{i=1}^{n} P(x_i)\log_2 \frac{1}{P(x_i)}$$

$$= - \sum_{i=1}^{n} P(x_i)\log_2 P(x_i) \text{ bits/message}.$$

The quantity H is called the 'source entropy', since it represents 'a degree of disorder' in the message selection at the source.

If all the probabilities $P(x_i)$ are equal, $P(x_i) = 1/n$ and $H = \log n$. H therefore increases with n. This is reasonable since the more messages there are to select from, the greater the uncertainty as to which message will be selected. There is therefore a greater change in information when the outcome of the selection is known. For a given value of n, H is a maximum and equal to $\log n$ when all the $P(x_i)$ are equal. This too is reasonable since this situation is associated with the greatest uncertainty of choice. Since for equiprobable events $H = \log n$, we can define 1 bit as the amount of information obtained when the selection is made from two equiprobable events, 1 hartley as the amount of information obtained when the selection is made for 10 equiprobable events, and 1 nat as the amount of information obtained when the selection is made from e equiprobable events.

It is worth noting that $H = 0$ if, and only if, all the $P(x_i)$ are zero except one, which is unity. That is, if a particular selection is certain to be made, then there is no information to be gained from actually making the selection. It is also worth noting that H can never be negative. Thus it is impossible to reduce the amount of information already known by making a further message selection At worst it can only give you no further information.

If an information source selects from a set of only two messages x_1 and x_2, then it is said to be a binary source. If the probabilities of the two messages occurring are P and $1-P$ respectively, then the entropy function of a zero-memory source is:

$$H = -P\log_2 P - (1-P)\log_2(1-P) .$$

This function is plotted in Fig. 5.1.

The symbol used to represent the output from a binary source is often

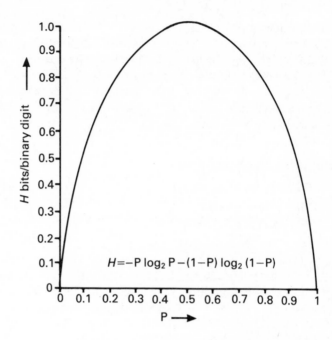

Fig. 5.1 – Entropy function for binary memory-less channel.

referred to as a binary bit. This must be distinguished from the bit which is used as a measure of information. It can be seen from Fig. 5.1 that on average the amount of information provided by a binary source is always equal or less than 1 bit/binary symbol. The binary source only provides one bit of information for each binary symbol generated when the two messages are equiprobable.

The entropy H of a source is thus defined as the average amount of information provided by the source per message generated. Alternatively, it can also be interpreted as the average amount of information necessary to specify which message has been generated. This means that, theoretically, any sequence of n mesages generated by the source can be encoded and transmitted using only nH binary symbols, each carrying one bit of information. We thus have the answer to the question of how many binary digits are necessary to specify unambiguously the output of a source that selects its output from an ensemble of messages according to some probability rule. The fact that the amount of information generated by a source is H and that this is exactly the minimum number of binary digits necessary to specify the output unequivocally is a good justification of the use of this information measure.

BINARY ENCODING AN INFORMATION SOURCE

When a message is selected by an information source an average amount of information equal to H is generated. Thus, on average, it should be possible to transmit the selected message using only H binary digits, H being the lower limit. Usually, in practice, many more digits are used than are theoretically necessary. We shall now consider two basically similar methods for encoding the output of the source so as to represent the message unambiguously using as few binary digits as is practically reasonable. The first of these methods is that known as Shannon–Fano encoding.

SHANNON–FANO ENCODING

In the Shannon–Fano encoding procedure we first arrange the messages in order of decreasing probability. The messages are then divided into two groups, the two groups having as nearly equal total probabilities as possible. The binary

Arrangement A

Source message	Probability $P(x_i)$	Code-words representing each measure	
x_1	0.4	0	Code-word 1
x_2	0.2	1 0	Code-word 2
x_3	0.2	1 1 0	Code-word 3
x_4	0.1	1 1 1 0	Code-word 4
x_5	0.07	1 1 1 1 0	Code-word 5
x_6	0.03	1 1 1 1 1	Code-word 6

Arrangement B

Source message	Probability $P(x_i)$	Code-words representing each measure	
x_1	0.4	0 0	Code-word 1
x_2	0.2	0 1	Code-word 2
x_3	0.2	1 0	Code-word 3
x_4	0.1	1 1 0	Code-word 4
x_5	0.07	1 1 1 0	Code-word 5
x_6	0.03	1 1 1 1	Code-word 6

Fig. 5.2 – Example of Shannon-Fano encoding.

symbol 0 is assigned to each message in the upper group and the binary symbol 1 to each message in the lower group as shown in Fig. 5.2. The process is then repeated by dividing each of the two groups into subgroups of nearly equal probability. A binary 0 is then assigned to each message in the upper subgroup of each group and a binary 1 to each message in the lower subgroup of each group. This process is continued until each subgroup contains only one message.

Note that a string of consecutively transmitted code-words can be unambiguously decoded, since no individual code word is the same as the beginning of another. Thus all that is required is that binary digits are accepted sequentially until a recognised code-word has been received.

It will be seen that the length of the code-words vary according to the probability of selection of the message. Frequently selected messages have few digits and infrequently selected messages have relatively more digits. We can calculate the source entropy and the average code-word length for each of the coding arrangements shown in Fig. 5.2.

The source entropy

$$H = -(0.4 \log_2 0.4 + 0.2 \log_2 0.2 + 0.2 \log_2 0.2 + 0.1 \log_2 0.1 + 0.07 \log_2 0.07 + 0.3 \log_2 0.3)$$

$$= 2.21 \text{ bits/message.}$$

For arrangement A:

Average word length $= 1 \times 0.4 + 2 \times 0.2 + 3 \times 0.2 + 4 \times 0.1 + 5 \times 0.07 + 5 \times 0.03$

$$= 2.3 \text{ binary digits/message.}$$

For arrangement B:

Average word length $= 2 \times 0.4 + 2 \times 0.2 + 2 \times 0.2 + 3 \times 0.1 + 4 \times 0.07 + 4 \times 0.03$

$$= 2.3 \text{ binary digits/message.}$$

This is the same as for arrangement A.

Since the source entropy represents the minimum average number of bits required to represent the message sequence unambiguously, we can define a coding efficiency as:

$$\text{Coding efficiency} = \frac{\text{source entropy}}{\text{average code-word length}} \times 100\%$$

For the example given, the coding efficiency is therefore

$$= \frac{2.21}{2.3} \times 100\% = 96\% .$$

HUFFMAN ENCODING

Although the Shannon–Fano method of encoding is generally satisfactory, there is no guarantee that the average number of binary digits used to represent a source message will be as small as, or smaller than, the average number of binary digits used when encoded by some other scheme. An encoding procedure, developed by Huffman, is optimum in the sense that on average, no other encoding scheme uses fewer binary digits to represent a message. The Huffman encoding scheme is as follows. Firstly the messages are arranged in order of decreasing probability as for the Shannon–Fano method. The two messages of lowest probability are then combined to form a single message whose probability is the sum of the two constituent messages. A new message set is then formed from the original set, with the combined message replacing its two con-

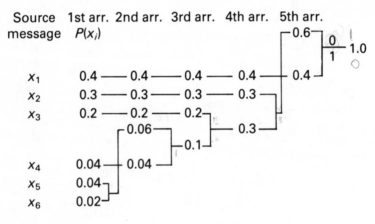

(a) Coding tree

Source message	Code-words
x_1	1
x_2	0 0
x_3	0 1 0
x_4	0 1 1 1
x_5	0 1 1 0 0
x_6	0 1 1 0 1

(b) Code-words

Fig. 5.3 – Example of Huffman coding.

stituent messages as shown in Fig. 5.3(a), the new message set being again arranged in order of decreasing probability. The procedure of combining the two messages of least probability and re-ordering is then repeated until a single message of unity probability is obtained. Wherever two messages have been combined to form a new message, a binary zero is assigned to the upper message in the combination and a binary óne to the lower message in the combination. The complete code-word for a particular source message is the sequence of binary digits leading from the final unity probability message back through the various message junctions to the source message in question. The code-words for the set of messages given in the example are shown in Fig. 5.3(b). Calculating the average code-word length and source entropy as before we have:

$$\text{average code-word length} \quad = 2.06 \quad \text{binary digits/message.}$$

$$\text{source entropy} \quad = 1.999 \quad \text{bits/message,}$$

$$\text{and hence the coding efficiency} \; \doteqdot \; \frac{1.999}{2.06} \times 100 = 97\%.$$

As with Shannon–Fano encoding, no complete code-word is the same sequence as the beginning of another code-word, so a stream of binary digits can be divided up into individual message codes as before.

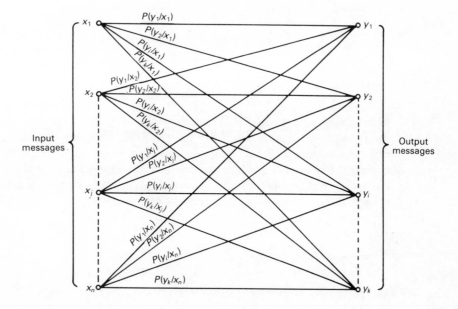

Fig. 5.4 – Channel representation using forward probabilities.

THE REPRESENTATION OF A CHANNEL

After a message has been selected from a source and suitably encoded, it is then fed to the communication channel for onward transmission. At the receiving end of the channel a decision is made as to the message that was transmitted and this constitutes the output from the channel. Because of various forms of signal impairment as, for example, noise interference, incorrect decisions will be made from time to time and the output of the channel will sometimes differ from its input. The decisions mde by the detector can thus be related to the channel input by a set of conditional probabilities.

If the set of n input messages are denoted by $x_1, x_2, \ldots, x_n$ and the set of k output messages by $y_1, y_2, \ldots, y_k$, then the channel, which includes the transmitter and the receiver, can be represented as shown in Fig. 5.4. Note that in general k does not necessarily have to equal n, though in most practical applications this is likely to be the case. Alternatively the channel can be represented by a channel matrix as follows:

$$
\begin{array}{cc}
 & \text{Output messages} \\
\text{Input messages} \left\{ \begin{array}{c} x_1 \\ x_2 \\ \vdots \\ x_j \\ \vdots \\ x_n \end{array} \right.
&
\begin{array}{ccccc}
y_1 & y_2 & y_i & & y_k \\
\left[\begin{array}{ccccc}
P(y_1/x_1) & P(y_2/x_1) & \cdots & P(y_i/x_1) & \cdots & P(y_k/x_1) \\
P(y_1/x_2) & P(y_2/x_2) & \cdots & P(y_i/x_2) & \cdots & P(y_k/x_2) \\
\vdots & \vdots & & \vdots & & \vdots \\
P(y_1/x_j) & P(y_2/x_j) & \cdots & P(y_i/x_j) & \cdots & P(y_k/x_j) \\
\vdots & \vdots & & \vdots & & \vdots \\
P(y_1/x_n) & P(y_2/x_n) & \cdots & P(y_i/x_n) & \cdots & P(y_k/x_n)
\end{array} \right]
\end{array}
\end{array}
$$

In the diagram and matrix representations of the channel, the $P(y_i/x_j)$s are called the forward probabilities since they represent the probability that the receiver will make a decision that will result in an output message y_i when, in fact, the transmitted message was x_j. it thus defines the channel as viewed from the standpoint of the sender. Clearly, since for a particular input message a decision of some sort must be reached as regards the output message

$$\sum_{i=1}^{k} P(y_i/x_j) = 1 .$$

The probability of obtaining a message y_i as output from the channel will be

$$P(y_i) = \sum_{j=1}^{n} P(x_j) . P(y_i/x_j)$$

It thus follows, from Bayes Rule[†], that the probability that a message x_j was transmitted, given that the output from the channel is y_i is

$$P(x_j/y_i) = \frac{P(y_i/x_j) \cdot P(x_j)}{P(y_i)}$$

and therefore

$$P(x_j/y_i) = \frac{P(y_i/x_j) \cdot P(x_j)}{\sum\limits_{j=1}^{n} P(x_j) \cdot P(y_i/x_j)}$$

The $P(x_j/y_i)$s are called the reverse or backward probabilities since they represent the probability that the sender transmitted the message x_j, when, in fact, the receiver has made a decision that has resulted in the output message y_i. It thus defines the channel as viewed from the standpoint of the recipient.

A MEASURE OF INFORMATION TRANSMITTED OVER A CHANNEL

Before an output is obtained from a communication channel, the probability that message x_j is the channel input is $P(x_j)$. The entropy associated with the input messages is, therefore:

$$H(X) = \sum\limits_{j=1}^{n} P(x_j) \log_2 \frac{1}{P(x_j)} \text{ bits/message.}$$

This *a priori* or source entropy can be interpreted, as shown earlier, as the average number of bits of information carried by an input message or as the average number of binary digits necessary to specify an input message.

After reception of an output message y_i, the probabilities associated with the input messages are now $P(x_1/y_i)$, $P(x_2/y_i)$, . . . $P(x_n/y_i)$ and the entropy associated with the set of input messages $x_1, x_2, . . ., x_n$ for the given y_i is

$$H(X/y_i) = \sum\limits_{j=1}^{n} P(x_j/y_i) \log_2 \frac{1}{P(x_j/y_i)} \text{ bits/message.}$$

[†]Bayes Rule states that $P(A/B) \cdot P(B) = P(B/A) \cdot P(A)$

hence $P(A/B) = \dfrac{P(B/A) \cdot P(A)}{P(B)}$

Taking the average over all possible output messages gives:

$$H(X/Y) = \sum_{i=1}^{k} P(y_i) . H(X/y_i)$$

$$= \sum_{i=1}^{k} P(y_i) \sum_{j=1}^{n} P(x_j/y_i) \log_2 \frac{1}{P(x_j/y_i)}$$

$$= \sum_{i=1}^{k} \sum_{j=1}^{n} P(y_i) . P(x_j/y_i) \log_2 \frac{1}{P(x_j/y_i)}$$

$$= \sum_{i=1}^{k} \sum_{j=1}^{n} P(y_i, x_j) \log_2 \frac{1}{P(x_j/y_i)} \text{ bits/message.}$$

$H(X/Y)$ is known as the *a posteriori* entropy, or equivocation, and can be interpreted as the average number of bits of information associated with an input message after a message has already been received at the output of the channel. It is thus also the average number of bits that are necessary in order to specify unequivocally the input message after a message has already been received at the output of the channel. $H(X/Y)$ is thus a measure of the uncertainty associated with the input after an output has been received. This uncertainty is caused by channel impairments such as noise.

The difference between the *a priori* and the *a posteriori* entropies, $I = H(X) - H(X/Y)$, is called the mutual information or transmission rate. It follows from the interpretations of $H(X)$ and $H(X/Y)$ that I is a measure of the amount of information gained by the recipient as a result of observing the message at the output of the channel.

It is worth noting that if there is no uncertainty as to which message was transmitted when an output message is received, then $H(X/Y)$ is zero and the information gained on reception of an output message is $H(X)$, the entropy of the source.

There are two other properties of mutual information I that are also worth noting. Firstly, the value of I must always be equal to or greater than zero. This means that the average amount of information received through a channel will be non-negative. Secondly, the only conditions under which $I = 0$ is when the channel input and channel output are statistically independent, that is, when $P(x_j, y_i) = P(x_j) . P(y_i)$ and hence $P(x_j/y_i) = P(x_j)$. This is a reasonable property since statistical independence between the channel input and the channel output implies that the recipient learns nothing of the channel input from a knowledge of the channel output.

It will be seen from the discussion above that the mutual information depends not only on the fixed conditional probabilities relating the channel output to the channel input, but also on the probabilities with which the various channel input messages are chosen. By a suitable matching process it is therefore possible to match the source output messages to the channel input messages in such a way as to maximise the mutual information for a fixed set of channel conditional probabilities. This process is usually referred to as statistical matching.

The maximum value of mutual information obtainable for a particular channel and message source is known as the channel capacity. It should be noted that as the channel capacity depends on the message source as well as the channel itself; changing the message source, or the message probabilities associated with the message source, will change the channel capacity. An important theorem for information theory arises from the concept of channel capacity, known as Shannon's Second Theorem or the Channel Coding Theorem. This theorem states that 'If an information source has an entropy H and a channel capacity C then, provided $H < C$, the output from the source can be transmitted over the channel and recovered with an arbitrarily small probability of error. If $H > C$ then it is not possible to transmit and recover information with an arbitrarily small probability to error'. Unfortunately, although the theorem tells us what is and what is not possible, it does not tell us how we can achieve error-free transmission. In fact, rather complex coding procedures are normally involved to get anywhere near the optimum transmission rate for a channel. Nevertheless, the theorem does give us an indication as to whether or not we are wasting our time trying to overcome a fundamental limitation.

THE BINARY SYMMETRIC CHANNEL

The binary symmetric channel is of special interest because it is one which is frequently encountered in practice in digital transmission systems. The channel source is binary and the two binary states occur with equal probability. The channel forward probabilities are symmetrical as shown in Fig. 5.5. The mutual information, and hence the channel capacity C, is given by:

$$C = 1 + p \log_2 p + (1-p) \log_2 (1-p) \text{ bits/message,}$$

where $p = P(y_1/x_1) = P(y_2/x_2)$ and hence $(1-p) = P(y_1/x_2) = P(y_2/x_1)$. Assum-

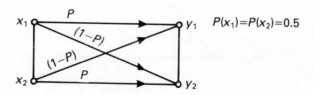

Fig. 5.5 – The binary symmetric channel.

ing a one-to-one mapping of the channel input messages and the desired channel output messages, we can then interpret p as the probability of correct transmission and $(1 - p)$ as the probability of error in transmission. If we assume the binary transmission is in the form of bipolar binary pulses and that errors in transmission occur because of impairment by additive white gaussian noise, then as is shown in the Appendix, we can express the probability of error in terms of signal-to-noise ratio S/N as follows:

$$p = \frac{1}{2} \operatorname{erfc}\left[\left(\frac{S}{2N}\right)^{1/2}\right]$$

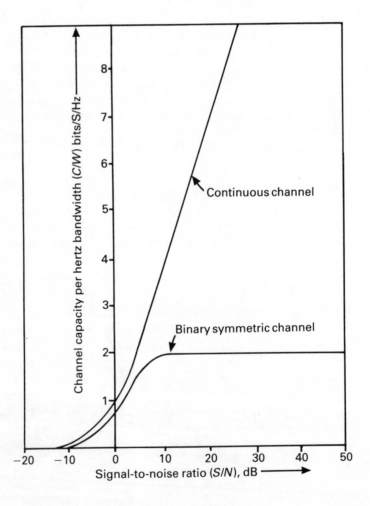

Fig. 5.6 – Channel capacity with additive Gaussian noise.

The channel capacity of a binary symmetric channel impaired by additive white gaussian noise is plotted in Fig. 5.6. It is plotted alongside the channel capacity for a continuous channel with the same impairment for the purpose of comparison. It will be seen that the maximum capacity for a binary symmetric channel

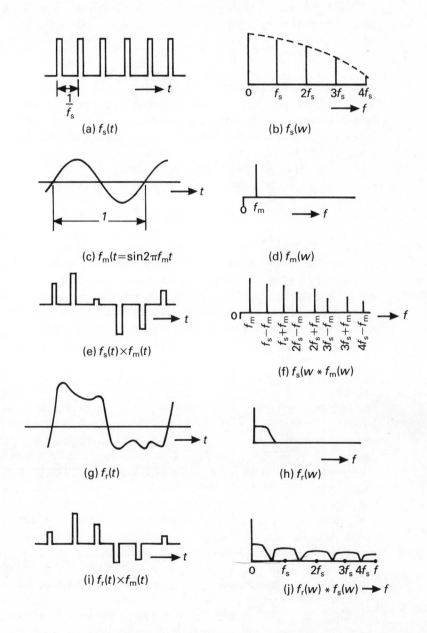

Fig. 5.7 – Signals and spectea of sampled waveforms.

is two bits per hertz of channel bandwidth. This assumes that pulses can be transmitted at a rate equal to twice the bandwidth in hertz. The validity of this assumption will be demonstrated in the next chapter. The maximum information that can be conveyed per binary symbol impulse is thus one bit. This occurs at very high signal-to-noise ratios and as the noise increases the information conveyed per binary symbol impulse decreases until, in the worst case, it reaches zero.

THE CONTINUOUS INFORMATION SOURCE

We can extend the theory we have developed for discrete information sources to cover the case of the continuous information source. However, in this case we have an infinite number of possible source messages, each of which has a very low probability of occurrence. We have therefore instead to consider continuous signals in terms of probability distributions. Before we proceed further, we need to look briefly at the sampling theorem, since this will prove to be a useful tool in our study of continuous signals.

THE SAMPLING THEOREM

It is possible, providing sufficient samples are taken, to fully represent any waveform by a series of samples which represent the amplitude of the signal at regular intervals in time. The sampling theorem tells us the minimum rate at which a particular waveform must be sampled to fully recover the initial waveform from the signal samples. We can consider the sampling waveform as a series of impulses $f_s(t)$ as shown in Fig. 5.7(a). Let us assume, in the first instance, that the waveform to be sampled consists of a single sinusoid of frequency f_m as shown in Fig. 5.7(c). The sampling operation can be regarded as the product of the sampling and sampled waveforms as shown in Fig. 5.7(e). The spectrum of the sampling waveform $f_s(t)$ consists of a series of spectral lines spaced at a frequency interval of f_s as shown in Fig. 5.7(b), f_s being the reciprocal of the sampling interval as shown in Fig. 5.7(a). Each component of the sampling signal spectrum will be amplitude-modulated by the multiplication operation so that it has side-bands spaced $\pm f_m$ from the component. This yields a sampled signal spectrum as shown in Fig. 5.7(f). If the sampled signal is replaced by a complex waveform having a spectral envelope as illustrated in Fig. 5.7(h), then the result of sampling will be a signal having a spectrum as illustrated in Fig. 5.7(j). Thus the effect of sampling is to produce double-sided spectral lobes about each harmonic of the sampling frequency f_s.

To recover a sampled signal we can filter off all the higher frequency components produced by sampling to leave simply the base-band spectrum as shown in Fig. 5.8(a). If the upper frequency in the base-band signal is increased, or the sampling frequency is decreased, the upper extremity of the base-band spectrum moves towards the lower extremity of the lower side-band associated with the fundamental of the sampling frequency. As these coalesce, the base-band extraction filter becomes more critical in its design as shown in Fig. 5.8(b). Further increase in base-band spectrum, or reduction of sampling frequency, will result in an overlap of the two spectrum lobes and proper separation becomes impossible, as shown in Fig. 5.8(c). This places a constraint on the choice of sampling frequency in as much as the sampling frequency must be equal to, or greater than, twice the highest frequency present in the base-band signal. If not, then the recovered signal becomes distorted owing to frequency components of the lower side-band associated with the sampling frequency fundamental becoming translated into the base-band spectrum. These frequencies experience a frequency translation which causes serious distortion of the base-band signal. This frequency translation is known as aliasing. The constraint that the sampling frequency must be at least twice the highest frequency present in the base-band signal is known as the sampling theorem.

THE ENTROPY OF A CONTINUOUS SIGNAL SOURCE

By regarding the continuous source as a limiting case of the discrete source, we can define the entropy of a continuous variable x with probability density function $p(x)$ as

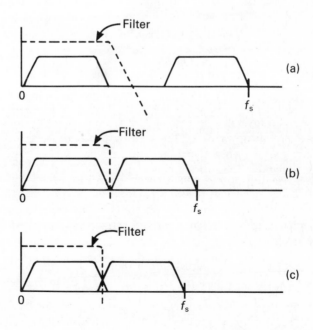

Fig. 5.8 – Recovery of sampled signals.

$$H(x) = \int_{-\infty}^{\infty} p(x) \log_2 p(x) \, dx \text{ bits/sample.}$$

If x is a continuous variable with probability density function $p(x)$ and variance σ^2, the form of $p(x)$ which gives the maximum entropy is the gaussian distribution. Thus $H(x)$ is maximum if

$$p(x) = \frac{1}{\sqrt{2\pi}\sigma} e^{-(x^2/2\sigma^2)}$$

This gives a value of entropy

$$H(x) = \log_2 \sqrt{2\pi e \sigma^2} \text{ bits/sample.}$$

If we sample a signal at the minimum sampling rate, then the sample values will be independent. If we sample below this rate, then we lose information about the signal. If we over-sample we gain no further information about the signal since the samples are no longer statistically independent. The entropy per second of a sampled waveform having white gaussian noiselike characteristics is therefore

$$H = 2W \log_2 \sqrt{2\pi e \sigma^2} = W \log_2 2\pi e \sigma^2 \text{ bits/second,}$$

where W is the bandwidth in hertz of the noiselike signal. For a given signal power σ^2, this is the maximum possible value of entropy for any form of signal probability distribution. A non-gaussian signal of the same power will have a lower entropy value. In such cases it is common to specify such signals in terms of their 'entropy power'. The entropy power of a signal is the power of the white gaussian noise signal that will have the same entropy as the signal under consideration. The entropy power of any signal is thus less than, or at best equal to, its actual power.

THE MUTUAL INFORMATION AND CAPACITY OF A CONTINUOUS CHANNEL

We can define the mutual information or transmission rate of a continuous channel in a way similar to that used for the discrete channel, that is:

$$I = H(x) - H(x/y) \ .$$

Unfortunately it is very difficult to define the channel in terms of $H(x/y)$ because of the infinite variability of both x and y. However, by symmetry of the equation it can readily be shown that

$$H(x) - H(x/y) = H(y) - H(y/x) \ .$$

The output y of a continuous channel is related to the input x by $y = x + n$, where n is the noise. We can assume, for all practical purposes, that x and n are statistically independent. $H(y/x)$ can be interpreted as the entropy associated with the output signal when the input signal is known to the receiver. It is therefore the entropy associated with the noise. The mutual information of a continuous channel is therefore given by

$$I = H(y) - H(n) \ ,$$

where $H(n)$ is the entropy of the channel noise. The channel capacity is the maximum value of I obtainable, as with the discrete signal case.

The capacity of a continuous channel can be maximised by ensuring that the signal has a gaussian distribution, that is it consists of a signal with noise-like qualities. Obviously the capacity is highest when the noise entropy is zero, that is there is no noise. However, the noise is a feature of the channel and cannot normally be altered to maximise the mutual information. Since it is reasonable to model additive channel noise on the basis of a white gaussian noise model, it is useful to calculate the channel capacity for a continuous channel impaired by additive white guassian noise. If the noise is white gaussian then the entropy is given by:

$$H(n) = W \log_2 2\pi eN \text{ bits/second} \ ,$$

where N is the average noise power.

If the average received signal power is equal to S, then the total average received power is $S + N$. The signal entropy will be a maximum when this signal has a gaussian probability distribution when:

$$H(y) = W \log_2 2\pi e(S + N) \text{ bits/second} \ .$$

The channel capacity is then given by:

$$C = W \log_2 2\pi e(S + N) - W \log_2 2\pi eN$$

$$= W \log_2 (1 + S/N) \text{ bits/second} \ .$$

This equation is usually referred to as 'Shannon's equation'.

It should be emphasised at this point that this equation gives the capacity of a continuous channel whose output can take an infinite number of different amplitude values. It is not the capacity of a binary or multilevel, but finite, signal channel. The capacity of the continuous channel is plotted, together with that of the binary symmetric channel, in Fig. 5.6. A multilevel signalling channel will have a capacity that lies somewhere between these two limiting cases.

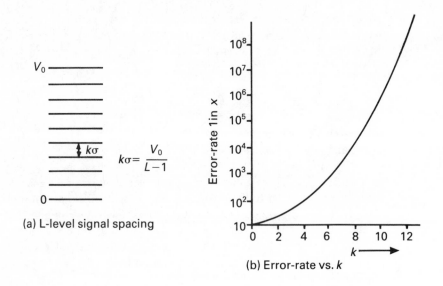

(a) L-level signal spacing

$k\sigma = \dfrac{V_0}{L-1}$

(b) Error-rate vs. k

Fig. 5.9 – Multilevel signalling by pulse-amplitude modulation.

It is of interest to digress at this point and briefly consider again multi-level pulse-amplitude-modulated transmission systems. Consider a system having L permissible signal amplitude levels as observed at the receiving end of the communication channel as shown in Fig. 5.9(a). Let $k\sigma$ be the voltage separation between levels, where σ is the rms value of the noise voltage. Clearly the larger the value of k, the smaller will be the probability that the noise will cause an error in deciding which level has been transmitted. If we assume the amplitude levels are from zero volts to a maximum of V_0 volts, then $V_0 = (L-1)k\sigma$. If we assume each amplitude is equiprobable, for maximum entropy, the average signal power

$$S_0 = k^2\sigma^2 \frac{1}{L} \sum_{a=0}^{(L-1)} a^2 \ .$$

Since the sum of the squares of the integer series from 0 to n is given by

$$\sum_{a=0}^{n} a^2 = \frac{n(n+1)(2n+1)}{6} \quad ,$$

the signal power

$$S_0 = \frac{k^2\sigma^2(L-1)(2L-1)}{6} \quad .$$

In fact the minimum average signal power for L equally spaced levels of interval $k\sigma$ will occur when the amplitude range is symmetrical about zero in the range $\frac{-k\sigma(L-1)}{2}$ to $\frac{+k\sigma(L-1)}{2}$. Thus the minimum average signal power

$$S = S_0 - \left(\frac{k\sigma(L-1)}{2}\right)^2$$

$$= k^2\sigma^2 \left[\frac{(L-1)(2L-1)}{6} - \frac{(L-1)^2}{4}\right]$$

$$= k^2\sigma^2\frac{L^2-1}{12} \quad .$$

However, the noise power $N = \sigma^2$, thus

$$S = k^2N\frac{L^2-1}{12}$$

and hence $L = \sqrt{\frac{12S}{k^2N} + 1}$ (5.1)

An L amplitude level signal is equivalent in information content to $\log_2 L$ binary digits. Thus the maximum transmission rate R_{max} possible with an L-level signal is given by:

$$R_{max} = 2W\log_2 L$$

$$= 2W\log_2\left(1 + \frac{12\,S}{k^2\,N}\right)^{1/2}$$

$$= W\log_2\left(1 + \frac{12\,S}{k^2\,N}\right)$$ (5.2)

Note how similar this is to Shannon's equation for the capacity of a continuous channel impaired by gaussian noise. A value of $k = \sqrt{12}$ would give the identical equation. However, because the signal does not have a gaussian distribution, the channel capacity will be less than for the continuous channel and some errors can therefore be expected for this value of k. In fact we can expect to get an error rate of about 1 in 25, that is, a probability of error of about 0.04. A value of k of 10 gives an error rate of about 1 in 10^6, or a probability of error of about 10^{-6}. The actual calculation of error probability is given in the Appendix. However, for interest, the error probability to be expected for various values of k, assuming the noise is white gaussian, is given in Fig. 5.9(b).

From equation (5.2) and from Shannon's equation we can see that, for a given information transmission rate, it is possible to make a trade-off between channel bandwidth and signal-to-noise- ratio. As the channel noise increases, the bandwidth can be increased to provide the same information transmission rate. In the case of multilevel signals, the increase in noise N will result in a decrease in the number of possible levels L for a given error-rate, determined by k. This can be clearly seen from equation (5.1). This principle will be of some importance to us when we consider the use of pulse-code-modulation in Chapter 7.

ERROR DETECTING AND CORRECTING CODES

We have seen how coding techniques can be used to minimise the redundancy in the signal content so as to make the most efficient use of the channel capacity. This type of coding is often referred to as 'coding down' because it is used to reduce to a minimum the number of bits of information that have to be transmitted in order to convey the required messages with an acceptable error-rate. We have also seen how errors can occur because of various channel impairments. By introducing extra 'redundant' bits into the data signal, it is possible to arrange for errors in the received data to be detected and, possibly, also corrected. The bits are redundant in the sense that they do not, of themselves, add anything to the information carrying capacity of the communication channel. However, they do carry information which enables the receiver to detect the presence of errors in the received data stream. The amount of redundancy needed depends on the proportion of errors acceptable after the operation of the decoding procedure and on whether it is necessary to locate errors so that they can be corrected or whether a general indication that there is an error in a message block is sufficient. Error-correcting codes require considerably more redundancy than those designed simply to detect the presence of errors. Coding designed for error detection and/or correction is often referred to as 'coding up' because of the addition of redundancy to the data sequence. We shall restrict our study of error detecting and correcting codes to binary data sequences since these have by far the widest application.

ERROR DETECTION BY PARITY

The simplest-error-detecting codes are those based on the block principle, where the data stream is divided into blocks and check bits are associated with each block transmitted. Generally speaking, the larger the block size the fewer the redundant bits that are required but the greater the probability of an error going undetected. The simplest method for detecting all single errors per block is the addition of a single 'parity' bit per block. The additional bit is chosen so that there is either always an even number of 1 bits per block, known as even parity, or an odd number of 1 bits per block, known as odd parity. An example of even parity is given in Fig. 5.10. On receipt, the parity of the block is checked and any violation of the parity rule will indicate an error somewhere in the block. Whether a 1 has been read as a 0, or a 0 as a 1, in either case the parity check will fail. Double errors will, of course, go undetected. The block size must therefore be chosen small enough to reduce the possibility of a double error occurring in a single block to an acceptable level.

Block code	Even parity
1 0 1 1 0 1 0 1	1
0 0 0 1 1 0 1 1	0

Fig. 5.10 – Example of even parity.

The concept of parity checking can be extended to detect and correct all single errors in a data block. The data block is arranged in rectangular matrix form as shown in Fig. 5.11. Parity bits are then provided for each row and each column in the matrix. A change in any single data bit will now cause the failure of two parity checks and the position of the parity failures will enable the

a	b	c	d	P_1
e	f	g	h	P_2
i	j	k	l	P_3
P_7	P_6	P_5	P_4	

Fig. 5.11 – Block parity for single error correction.

erroneous data bit to be located and hence corrected. Since the bit can only have one of two states, correction of an erroneous bit simply involves a reversal of state of that particular bit. An erroneous parity bit, and these are just as vulnerable as the data bits, will be indicated by one parity failure only.

Multiple errors, of course, will lead to meaningless or even misleading indications of parity failure. However, since the probability of single error is so much greater than that of a double error, and higher order errors are even less probable, the code can be used effectively to substantially reduce the error rate, provided the block size is chosen small enough. Note, however, that the redundancy increases rapidly with reduction in block size.

HAMMING DISTANCE

It is time we looked a little closer at the theory of error-detecting and -correcting codes in order to be able to specify the code capability. To simplify our terminology we shall use the concept of a code-word. A code-word is a group of binary digits taken together to form a recognisable coding entity. Thus a data block, together with the added redundant bits, may be regarded as a code-word. Hamming distance can then be defined as the number of corresponding bits in two code-words from a code-word group that are different in state. Thus the Hamming distance between the two words given below will be 3.

```
0   0   1   0   1   1   0   1
X               X       X
1   0   1   0   0   1   1   1
```

Similarly, the distance between the two following words will be 4.

```
0   0   0   0   1   1   1   1
        X   X   X   X
0   0   1   1   0   0   1   1
```

From the concept of Hamming distance we can deduce the following rules:

(a) If the minimum distance between any two words in a set of code-words is 1, then a single error is likely to change the word into another word in the set and error detection is thus impossible.

(b) If the minimum distance between any two words in a set of code-words is 2, then it is possible to detect all single errors since at least two errors are needed to alter any word into another in the set.

(c) If the minimum distance between any two words in a set of code-words is 3, then it is possible to correct all single errors since a single error will still leave the word at least two bits different from any other word in the set. It is therefore possible to determine which word was sent since the correct word will be the only one which has a distance of 1 from the received word. Unfortunately, a double error is likely to lead to a third error being produced and the three errors being passed on undetected, since two errors can result in a received word which differs in only one position from another word in the set. This would thus be interpreted as a single error which would be incorrectly interpreted as a code-word of distance 3 from the one transmitted. Code-words with a distance of 3 can be used to detect all single and double errors, providing single errors are not to be corrected.

(d) With a minimum distance of 4, it is possible to correct single errors and detect double errors at the same time. To correct both single and double errors we need a minimum distance of 5 between code-words. It can be shown that to correct up to x errors per code-word will require a Hamming distance of $D = 2x + 1$.

It is surprising how many redundant bits are required to obtain a Hamming distance of any significance. For example, it is only possible to select 4 code-words of 5 bit length having a Hamming distance of 3, that is, sufficient distance to carry out single error correction. The full set of words and their implication, for a particular selection of 4 words of 5 bits, is given in Fig. 5.12.

```
                    00000 ⎞
                    10011 ⎟   These are not the only·set of
                    11100 ⎬   four we could have chosen
                    01111 ⎠
     00000   10011   11100   01111        Code-word
     00001   10010   11101   01110
     00010   10001   11110   01101 ⎞
     00100   10111   11000   01011 ⎬   Single-bit errors
     01000   11011   10100   00111 ⎠
     10000   00011   01100   11111
   These combinations occur only with two errors:
     00101   00110   01001   01010
     11010   11001   10110   10101
```

Fig. 5.12 – Five bit code words with distance three.

HAMMING CODES

It is possible to devise effective single-error-correcting codes which are more efficient than the block parity check codes in that they require fewer redundant bits. It can be shown that optimum use of redundancy requires n bits of a block of size N to be parity bits, where $n = \log_2(N + 1)$. Since it is impossible to have fractional bits, n must be rounded up to the nearest integer. The most efficient block sizes are those that require no rounding up, that is, where $N = 2^n - 1$, n integer. The next bit to be added to a block of this size will have of necessity to be a parity check bit. In fact a systematic arrangement for such a code is for the bits in positions 2^n, $n = 0, 1, 2, 3, \ldots$, to be allocated for parity checking. A code arranged on this basis is generally referred to as a Hamming code. An example of a Hamming code for $N = 7, n = 3$, is given in Table 5.1. It will be evident from the table how the code can be extended for any required block size. The position of the error is indicated by parity check failures as shown in the table. The number of the position is given directly in binary form by entering 1 for a failure and 0 for a pass in the error check table. Errors are corrected by bit reversal. Note that errors in the parity bits are signified by a single indicated failure in the check table.

Table 5.1 – Hamming code for $N = 7, n = 3$

	K_1	K_2	Data	K_3		Data	
Data	1	2	3	4	5	6	7
0	0	0	0	0	0	0	0
1	1	1	0	1	0	0	1
2	0	1	0	1	0	1	0
3	1	0	0	0	0	1	1
4	1	0	0	1	1	0	0
5	0	1	0	0	1	0	1
6	1	1	0	0	1	1	0
7	0	0	0	1	1	1	1
8	1	1	1	0	0	0	0
9	0	0	1	1	0	0	1
10	1	0	1	1	0	1	0
11	0	1	1	0	0	1	1
12	0	1	1	1	1	0	0
13	1	0	1	0	1	0	1
14	0	0	1	0	1	1	0
15	1	1	1	1	1	1	1

K_1 is even parity on positions 1, 3, 5, 7.
K_2 is even parity on positions 2, 3, 6, 7.
K_3 is even parity on positions 4, 5, 6, 7.

Data is given in binary form in positions 3, 5, 6, 7.

Single errors cause failure of parity checks thus:
$$(1 = \text{failure}, 0 = \text{check passed})$$

	K_3	K_2	K_1
No error	0	0	0
Error in position 1	0	0	1
Error in position 2	0	1	0
Error in position 3	0	1	1
Error in position 4	1	0	0
Error in position 5	1	0	1
Error in position 6	1	1	0
Error in position 7	1	1	1

It will now be evident why it was only possible to select four code-words with minimum distance 3 from the set of 5-bit code-words. In this case bits 1, 2 and 4 must be parity check bits, leaving only bits 3 and 5 for data. This permits only four data combinations. The four words selected in the example given in Fig. 5.12 are in fact those given by a 5-bit Hamming code selection.

It is, of course, not necessary that the bits be transmitted in the strict sequence of the coding rule. An alternative would be to transmit the data bits first, followed by the parity checks. However, the rules for determining the value of the parity bits and the position of the detected errors must be based on the basic code structure for satisfactory operation.

ALGEBRAIC CODES

The algebraic codes present a more general approach to error-correcting codes and include the Hamming code as a special case. Let us assume, as before, that our set of code-words consists of words of N digits, n of these digits being parity check digits. Thus $m = N - n$ digits actually carry information. Thus, although with N-digit words it is possible to select 2^m different code-words, in fact only 2m words are used. We can express the rule used for the selection of the subset of code-words used using Boolean matrix algebra. Thus the words are chosen to satisfy the equation

$$[H]\, T' \;=\; 0$$

where T' is the N column vector which is the transpose of T, the N row vector representing the selected code word, and $[H]$ is an $n \times N$ matrix where each column is unique and non-zero. Thus, for example, the Hamming code for $N = 7, n = 3$, is generated using the matrix

$$[H] \;=\; \begin{bmatrix} 1 & 0 & 1 & 0 & 1 & 0 & 1 \\ 0 & 1 & 1 & 0 & 0 & 1 & 1 \\ 0 & 0 & 0 & 1 & 1 & 1 & 1 \end{bmatrix}$$

It is usual, though not essential, to take the positions represented by the columns containing a single 1 only as the parity check digits in the code-words. Thus for the Hamming code for $N = 7$, the check digits are in the first, second, and fourth positions. It is a simple matter to re-order the matrix so that the check digits occur after the information carrying digits. For example:

$$[H] \;=\; \begin{bmatrix} 1 & 1 & 0 & 1 & 1 & 0 & 0 \\ 1 & 0 & 1 & 1 & 0 & 1 & 0 \\ 0 & 1 & 1 & 1 & 0 & 0 & 1 \end{bmatrix}.$$

When a code-word is received, the receiver determines the product of the received code-word with the matrix $[H]$. Since for all acceptable code words the product $[H]\, T' = 0$, if the product formed at the receiver is other than zero, we know that at least a single error must have occurred in transmission.

If T is the row vector representing the transmitted word and R is the row vector representing the received word, then we can define an error vector E which contains a 1 in each position in which an error has occurred.

Then

$$R \;=\; T + E$$

and hence, by simple tranpositions into column vectors

$$R' \;=\; T' + E'$$

where the addition is Boolean (modulo 2).

Now the product at the receiver

$$[H]R' = [H](T' + E')$$
$$= [H]T' + [H]E'$$

However, by definition, $[H]T' = 0$, hence

$$[H]R' = [H]E' .$$

Thus our multiplication operation at the receiver gives us directly the value $[H]E'$, usually referred to as the code syndrome. We can then identify the position of a single error in the received code word by selecting the position of the column in the matrix H which corresponds with the syndrome $[H]E'$. Since the digits in the received message have only two possible states, any indicated error can easily be corrected by simply reversing the state of the digit in question. It will now be clear why each column in the matrix $[H]$ needs to be unique and non-zero. If it were zero, then it would be indistinguishable from the 'all correct' indication.

Let us take an example. If we assume that the word 1 1 0 1 0 0 1 from the Hamming code set was transmitted but was received as 1 1 0 1 1 0 1, when the receiver product is formed we get

$$[H]R' = \begin{bmatrix} 1 & 0 & 1 & 0 & 1 & 0 & 1 \\ 0 & 1 & 1 & 0 & 0 & 1 & 1 \\ 0 & 0 & 0 & 1 & 1 & 1 & 1 \end{bmatrix} \cdot \begin{bmatrix} 1 \\ 1 \\ 0 \\ 1 \\ 1 \\ 0 \\ 1 \end{bmatrix} = \begin{bmatrix} 1 \\ 0 \\ 1 \end{bmatrix} .$$

The syndrome is seen to be equal to the fifth column of the matrix $[H]$, thus indicating an error in the fifth digit of the received code-word. Inversion of the appropriate digit is all that is necessary to correct the error. The concept of algebraic codes can be extended to the correction of multiple errors but further consideration of these possibilities is beyond the scope of this book.

CYCLIC REDUNDANCY CHECK CODES

The identifying feature of a cyclic code is the property that a cyclic, or end-about, shift of any code-word results in another code-word from the set. The attractiveness of the cyclic codes is that they can be implemented fairly easily using shift-register techniques. A code-word of length N with m information digits is referred to as an (N, m) cyclic code. The words of the code set are formed by Boolean division of a code word of length N, consisting of m information digits followed by $n = N - m$ zeros, by a binary divisor of length $n + 1$. The remainder following the division is then inserted in place of the n zeros following the m information digits. Because of the complementary nature of modulo 2 division, the transmitted code-word will be divisible without remainder by the initial dividing 'characteristic polynomial'. Any remainder after so dividing a received code-word will indicate an error in transmission. The power of the code to detect and/or to correct errors is determined by the relative magnitudes of the information and check fields m and n and also by the selection of the characteristic polynomial. Unfortunately there are no simple rules for determining the best polynomials for any particular circumstance. Thus anyone designing a cyclic redundancy check code must refer to the extensive literature of the subject.

A widely used cyclic redundancy code, specified in CCITT recommendation V41 is the (256, 240) code with the divisor 1 0 0 0 1 0 0 0 0 0 0 1 0 0 0 0 1. It is usual, and less confusing, to specify the divisor in the form of a characteristic polynomial. The characteristic polynomial for the V41 divisor is

$$x^{16} + x^{12} + x^5 + 1 \ .$$

The index to each term in the polynomial indicates the position of a 1 in the divisor, numbering the least significant position as 0. The V41 code will only permit error detection and is normally used in conjunction with ARQ (request repeat of blocks detected to be in error) operation. However, the divisor has been so chosen as to give a high degree of multiple error detection, especially errors occurring in short bursts.

CONVOLUTION CODES

All the codes described so far operate on a block basis and are most efficient in dealing with errors which occur at random. Unfortunately, in practice, errors frequently occur in bursts since noise impairments are often impulsive in nature rather than being in the form of additive, white, gaussian noise. The convolution codes are particularly useful for detecting and correcting burst errors. In the convolution codes, the data stream is not divided into blocks. Instead, parity checks are made on digits separated by several further digits and the check digits are inserted between the information digits elsewhere in the stream. The Hagelbarger code is a basic form of convolution code.

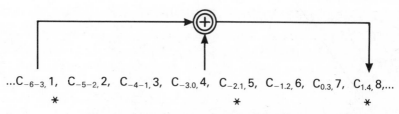

...$C_{-6,-3}$, 1, $C_{-5,-2}$, 2, $C_{-4,-1}$, 3, $C_{-3,0}$, 4, $C_{-2,1}$, 5, $C_{-1,2}$, 6, $C_{0,3}$, 7, $C_{1,4}$, 8,...

*Positions containing information relating to digit 1

Fig 5.13 – The Hagelbarger code.

THE HAGELBARGER CODE

In the Hagelbarger code alternate digits in the transmitted data stream are check digits formed by modulo-2 sum of the mth and nth preceding information digits. This means that information about a single digit is held in three places spread out in time according to the selection of the values of m and n. The redundancy of such a code is 50%. The Hagelbarger code is illustrated in Fig. 5.13. Note that in this example $m = 14$ and $n = 7$. In this particular example up to six consecutive errors would be corrected, provided there are at least 19 error-free digits between bursts. If the information relating to a single digit is spread more widely in the digit stream, then larger bursts can be corrected but longer error-free runs between bursts are necessary for satisfactory error correction.

ERROR CORRECTION BY RE-TRANSMISSION (ARQ)

The error-correcting codes we have discussed give what is called 'forward error correction', that is, the errors are corrected before the messages are passed to the receiver. The main disadvantage of forward error correction is the large amount of redundancy usually required for these codes. It is therefore often more convenient simply to detect errors and then request the re-transmission of any faulty blocks. This mode of operation is sometimes called ARQ. The practicability of doing this depends on the ability to transmit control information back to the sender and for the sender to store information until it has been acknowledged by the receiver. The process of acknowledgement is referred to as 'hand-shaking'. A disadvantage of ARQ is that, unless each block is stored at the receiver before being output to the terminal, it is not possible to have 'clean copy' output. The choice between forward error correction and ARQ depends on the availability of a return channel, the economics of implementing the coding and decoding strategies, the transmission time requirements and the probability and acceptability of transmission errors.

6

Data transmission

Data transmission is the process of communicating information in digital form from one location to another. If the communication is simply between two terminals in reasonably close geographical proximity, then there is really no problem in providing a dedicated connection with adequate bandwidth and suitably chosen characteristics to enable direct interconnection to be made. Difficulties arise, however, where the distances involved, and the usage is such, that it is uneconomic to provide dedicated special-purpose circuits. In this chapter we will consider the problems of long-distance data transmission.

THE BASIC PROBLEMS OF LONG-DISTANCE DATA TRANSMISSION

If data transmission systems consisted only of short lengths of properly matched coaxial cable, there would be little difficulty in digital data transmission. The output waveform from such a channel is virtually identical to the input waveform. However, when data is sent over networks which have not been designed with data transmission in mind, such as the public telephone network, the transmission path may differ from the ideal in the following ways.

(a) *Restricted bandwidth*. Frequency components beyond some upper limit are heavily attenuated by the channel. The effect of the limited high frequency response on a binary waveform is illustrated in Fig. 6.1(a).

(b) *No d.c. response*. Frequency components below some limit may be heavily attenuated, with no transmission at all at zero frequency. The effect of

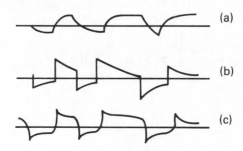

Fig. 6.1 – Effect of various degradations on binary data. (a) Limited h.f. response. (b) Limited l.f. response. (c) Phase distortion.

limited low frequency response on a binary waveform is illustrated in Fig. 6.1(b).

(c) *Gain-frequency and phrase-frequency distortion.* The gain of a channel will vary to some extent even between the lower and upper cut-off frequencies. This results in the output waveform differing from the input waveform because the frequency components have had their amplitudes changed by differing amounts. Similarly, if the phase-shift versus frequency characteristic of the channel deviates from a straight line (i.e. transmission delay varies with frequency), the frequency components are delayed by differing amounts, so that the output waveform differs from the input waveform. The effect of grain-frequency and phase-frequency distortion on a binary waveform is illustrated in Fig. 6.1(c).

(d) *Echoes.* Discontinuities in the transmission path cause reflections of the transmitted signals which occur at the receiver as signal echoes.

(e) *Additive gaussian noise.* Any transmission system incorporating amplifiers will introduce random noise, added to the signal waveform. In a telephone system, such noise can be heard as background hiss. This noise usually has zero d.c. value and an approximately gaussian amplitude distribution. This will cause a proportion of the symbols received to be in error. Note that, because the gaussian distribution tails off very rapidly, the error rate diminishes very rapidly indeed with increase in signal-to-noise ratio.

(f) *Burst noise.* Some transmission systems suffer from occasional bursts of high amplitude noise such as clicks and pops in telephone systems or ignition interference in radio systems. Normally all symbols occurring during such noise bursts are lost.

(g) *Other degradations.* The foregoing transmission impairments are the ones that normally give difficulty in practice. Other types of degradation exist and may give trouble in special circumstances. These include non-linear transmission characteristics due, for example, to amplifier overload, fre-

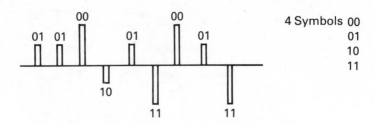

Fig. 6.2 – Data transmission through an ideal channel (2 bits per puese).

quency offsets in FDM systems and time-varying channel characteristics such as encountered in HF radio and underwater data transmission.

THE IDEAL CHANNEL

We saw in the previous chapter that the two transmission limitations that set a fundamental limit to the rate at which data can be transmitted are noise and restricted bandwidth. The ideal channel has constant gain from zero to infinite frequency, no phase-frequency distortion, no noise and no other transmission impairments. If we assume that transmission comprises short high amplitude pulses at intervals of T units of time, each pulse having one of L possible amplitude levels, then $\log_2 L$ bits of information may be conveyed by each pulse. Fig. 6.2 illustrates the transmitted waveform where four transmission levels are used (i.e. two bits of information per pulse). At the receiving end of the channel, the amplitude at each pulse sampling instant is measured to determine the symbol transmitted.

THE EFFECT OF THE CHANNEL ON THE DATA PULSE

When a data pulse is transmitted through a communication channel it is inevitably band-limited and distorted owing to non-linear amplitude and phase-versus-frequency characteristics. The transmitted and received pulses are normally defined as functions of time, whereas the channel characteristics are normally defined as functions of frequency. To calculate the effect of the channel on the pulse we therefore have to make use of the Fourier transform. The Fourier transform is a mathematical relationship between the time and frequency domain specification of responses and characteristics. There are two ways of using Fourier transforms to determine the effect of the channel on the pulse shape. These are illustrated in Fig. 6.3. Firstly, we could determine the spectral content of the input pulse $X(\omega)$. By multiplying this with the channel characteristic $G(\omega)$ we obtain the spectral content of the output pulse $Y(\omega)$, which we

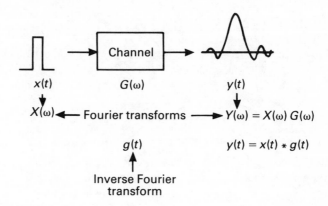

Fig. 6.3 – Use of Fourier transform to determine network response. The asterisk denotes convolution, i.e. $y(t) = \int_0^t x(\tau) \cdot g(t-\tau)d\tau$.

can now transform back into the time response $y(t)$. An alternative approach is to determine the inverse Fourier transform of he channel characteristic $g(t)$ and to convolve this with the input pulse response $x(t)$. This will yield the output pulse response $y(t)$ directly. The inverse Fourier transform of the channel characteristic $g(t)$ is the channel impulse response and is the output that would be obtained if the input to the channel were a 'dirac' impulse $\delta(t)$. This is illustrated in Fig. 6.4. The dirac impulse consists of an idealised pulse of infinite amplitude and infinitesimal time duration and having an 'area' of unity. It is, of course, impossible to generate such an impulse in practice, but it is a convenient mathematical concept for analysis of this kind. It should be noted that a network is uniquely defined by its impulse response in exactly the same way as it is defined by its amplitude and phase-versus-frequency characteristics.

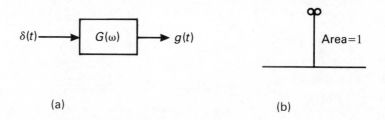

(a)   (b)

Fig. 6.4 – Definition of impulse response. (a) Impulse response $g(t)$. (b) Dirac impulse $\delta(t)$.

EFFECT OF BAND-LIMITING ON A DIRAC IMPULSE

If a channel has ideal characteristics except that frequencies beyond some 'cut-off' frequency are completely attenuated, the response of such a channel to an impulse is the familiar $(\sin x)/x$ form:

$$g(t) = \frac{\sin 2\pi Wt}{2\pi Wt}$$

Thus if a pulse which is short enough to be considered an impulse is applied to the input of an ideal band-limited channel, the waveform at the channel output is a $\sin x/x$ waveform with oscillating tails as shown in Fig. 6.5. If we just wanted

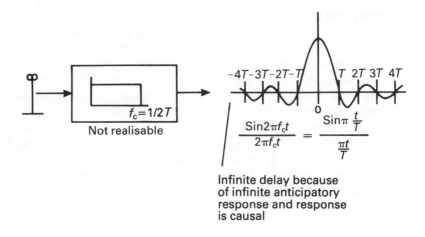

Fig. 6.5 – Impulse response of ideal band-limited channel.

to transmit a single pulse, there would be no problem, but when we transmit a succession of pulses, the tails caused by previous pulses may obscure the main response due to the pulse of the present instant. This is termed 'intersymbol interference'. Intersymbol interference can be completely avoided in the ideal band-limited channel by sending pulses with a pulse spacing of $T = 1/2W$. That is, if the cut-off frequency of the channel is W Hz, we send $2W$ pulses per second. This is because $g(t)$, the channel impulses response, is zero at $\pm 1/2W$, $\pm 2/2W$, $\pm 3/2W$, etc. Thus, at $t = nT$, the channel output is solely the main lobe of the response of the channel to the nth input pulse; the response of the channel to all the other input pulses is zero at this instant, as shown in Fig. 6.6.

In practice it is difficult to construct band-limiting filters giving a good approximation to the $(\sin x)/x$ impulse response and, in any case, a system using

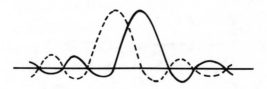

Fig. 6.6 – Channel response with zero inter-symbol interference at sampling instants.

the $(\sin x)/x$ response would be sensitive to small timing errors. What is needed is a channel whose response has the same property as the $(\sin x)/x$ function in being zero at $t = \pm T$, $\pm 2T$, etc. but which has a gentle roll-off with frequency. Nyquist shows that if the channel gain characteristic has 'vestigial' symmetry about a frequency equal to half the pulse transmission rate, and has a linear phase characteristic, then the impulse response of the channel has nulls at the appropriate points. Vestigial symmetry implies odd symmetry but offset so that the characteristic subsequently falls to zero value rather than ranging between equal positive and negative values. A suitable response which can be closely approximated to in practice is that known as the 'raised cosine roll-off' characteristic. This is illustrated in Fig. 6.7(a). The impulse responses corresponding to various roll-off factors are shown in Fig. 6.7(b). It will be seen that the oscillations in the pulse tails decrease as the excess bandwidth is increased.

MODIFICATION OF CHANNEL CHARACTERISTIC FOR FINITE DURATION PULSES

In practice we shall not normally be transmitting impulses, but finite duration pulses of time duration T. If we take a train of pulses of finite duration T repeated at intervals of time T as shown in Fig. 6.8(b), then, by Fourier analysis, we obtain the spectrum of the pulse train as illustrated in Fig. 6.8(a). The spectrum consists of lines spaced at frequency intervals of $1/T$ Hz, the spectrum envelope following the $\sin x/x$ shape, with the zeros in the $\sin x/x$ function occurring at frequencies n/T Hz, where $n = \pm 1$, ± 2, ± 3, etc. Clearly, as the interval between the pulses is increased, the spectrum lines close up and in the limit, when the second pulse goes to infinity and we are left with one single pulse, the spectrum becomes continuous with a $\sin x/x$ envelope. Similarly, if the pulse duration is reduced, the zeros in the $\sin x/x$ envelope move apart and in the limit, when the pulse becomes a dirac impulse, the first zero moves out to infinity and the spectrum becomes flat over any finite range. Thus the spectrum of a single dirac impulse is continuous and flat over the whole frequency range. The desired channel response to a finite duration pulse is the same as the impulse response where the vestigial symmetry of the channel occurs

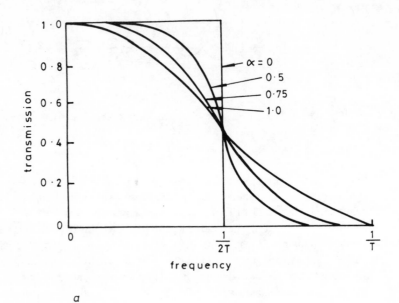

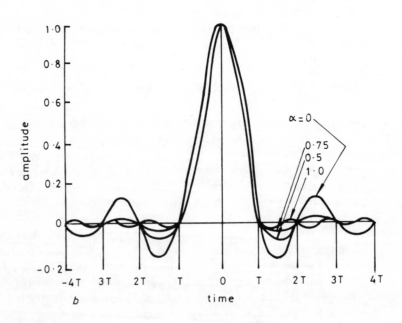

Fig. 6.7 – The raised cosine filter characteristics. (a) Raised-cosine roll-off characteristic. (b) Impulse responses of r.c.r.o. characteristics.

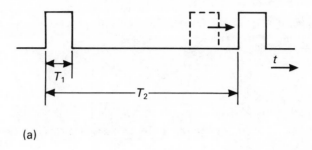

(a)

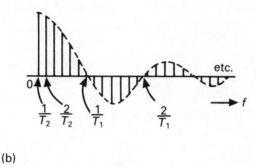

(b)

Fig. 6.8 — Pulse spectrum. (a) Repetitive pulse waveform. (b) Repetitive pulse spectrum.

about a frequency $1/2T$. To achieve this we need to modify the channel characteristic to allow for the $(\sin x)/x$ spectrum shape of the finite duration pulse. The concept is illustrated in Fig. 6.9. Since the spectrum for the impulse is flat, $F(\omega)$ is the inverse of the spectrum $X(\omega)$ for the pulse $x(t)$, i.e.

$$F(\omega) = \frac{1}{X(\omega)} \ .$$

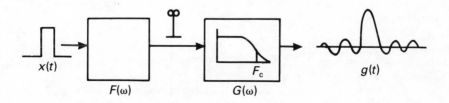

Fig. 6.9 — Allowance for finite duration pulse.

Thus, if we wish to obtain $g(t)$ from $x(t)$, we need to pass the signal through a channel with a gain-frequency response

$$Y(\omega) = \frac{1}{X(\omega)} \times G(\omega) \ .$$

Since $G(\omega) = 0$ for $f \geqslant f_c + f_r$, we need not define $X(\omega)$ outside this range. In this way we can determine the channel response required to give zero inter-

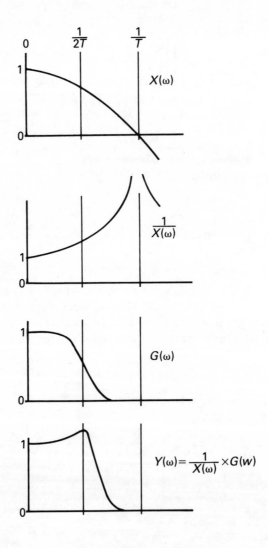

Fig. 6.10 – Ideal channel response for finite duration pulse.

symbol interference at sampling instants spaced at intervals of T time units apart for data represented by finite-width pulses of duration T. The process is illustrated in Fig. 6.10.

EYE PATTERNS

The eye pattern is a convenient method of displaying on an oscilloscope the

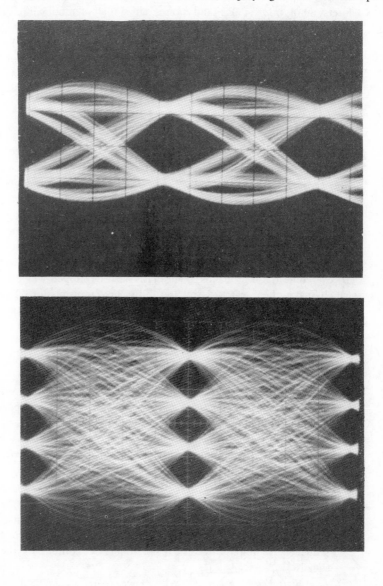

Fig. 6.11 – Eye patterns. (a) Binary eye pattern. (b) Four-level eye pattern.

effect of noise and inter-symbol interference on a received data signal. The received signal voltage is used to deflect the oscilloscope trace vertically and the horizontal sweep is synchronised to the data symbol rate. The resulting display for a random bipolar binary data sequence is shown in Fig. 6.11(a). The reason for the name 'eye pattern' is apparent. The eye pattern for a four-level random data sequence is shown in Fig. 6.11(b). The optimum sampling instant to ensure correct decoding is where the 'eye' is widest open. In the absence of intersymbol interference and noise, all similar symbol amplitudes will pass through the same point at the optimum sampling instant. The effect of the tails in the response of adjacent symbols becomes significant away from this point and the noise margin before the onset of errors is therefore significantly reduced if the signal is not optimally sampled.

LINE CODES

Neither the simple bipolar binary code, where 1 is represented by a positive pulse and 0 is represented by a negative pulse, nor the simple unipolar binary code, where 1 is represented by a positive pulse and 0 is represented by no signal, is ideal for line transmission purposes. Both suffer from the existence of strong frequency components at low frequencies and a lack of signal transitions when long strings of identical symbols are transmitted sequentially. These transitions are necessary to derive timing singals for the receiver decoder to enable the decoder to determine the number of consecutive bits in the sequence. To overcome these problems, codes have been developed which have properties which make them particularly useful for line transmission. These codes are usually referred to as line codes and it is usual to convert the signal to be transmitted into a line code for transmission purposes. The desirable features of a line code are as follows:

(a) *Transparency*. The code must not impose any restriction on the content of the transmitted message, i.e. it must be bit-sequence-independent.
(b) *Unique decodability*. Each output symbol must be unambiguously decoded to give the original sequence of input bits.
(c) *Efficiency*. Each symbol of the code should contribute to the transmission of the incoming information.
(d) *Favourable energy spectrum*. The presence of transformer and coupling capacitors in the system requires the line code to have a d.c. component averaging zero and very small low-frequency components. Since inter-channel interference increases with frequency, interference power can be reduced by minimising the high frequency content of the transmitted signals.
(e) *Low digital sum variation (DSV)*. If we define a running digital sum (RDS) as

$$RDS(k) = \sum_{n=1}^{k} C_n + RDS(0) ,$$

where $C_n = 1, 0$ or -1 for a ternary signal (or ±1 for binary), and $RDS(0)$ is an appropriately chosen constant, the DSV is then given by

$$RDS(\text{Max}) - RDS(\text{Min}) .$$

This denotes an upper bound on the string of like pulses and is roughly proportional to the peak-to-peak value of the low-frequency components. Those codes with the lowest DSV are therefore to be preferred. The DSV is often referred to as the code disparity.

(f) *Timing information content.* It is desirable that the spectrum of the transmitted signal has a high energy content close to the data clock frequency. This will be obtained if the line signal has frequent transitions at the clock interval.

Line codes for binary data transmission generally divide into two types. The first type are still binary in nature, but the code structure is modified to improve the code properties. Two such codes are the Walsh function codes Wal 1 and Wal 2. These are illustrated in Fig. 6.12(a). In both cases the line signal is free from d.c. component and contains a large number of transitions from which timing information can be recovered, whatever the transmitted data pattern. The line spectra for these two codes are shown in Fig. 612(b), where they are compared with the spectrum for the simple bipolar binary code. In both cases the spectrum has no d.c. component and low l.f. components. The bandwidth occupancy of the Wal 2 is slightly greater than the Wal 1, extending to approximately twice the transmission rate. The Wal 2 spectrum has no significant content below 0.2 of the transmission rate, a fact that can be made use of in data-over-voice applications. Another useful binary code is the Miller code, or delay modulation, as it is sometimes called. This code is a variation of the Wal 1 and is derived by deleting every second transition in a Wal 1 signal, as illustrated in Fig. 6.12(a) — The spectrum shape of the line signal is given in Fig. 6.12(b), where it can be compared with the Wal 1 and Wal 2 spectra. Although it has a small d.c. component, it has the advantage of a more limited bandwidth requirement than the comparable Wal codes. The Wal 1 code is sometimes referred to as Manchester encoding.

The second type of line codes are ternary codes which operate on three signal levels, the middle level of which is usually zero volts. These codes are sometimes referred to as pseudo-ternary since, although the code has three levels and therefore a capability greater than binary coding, each symbol only conveys one bit of information. The best known of the tenary codes is alternate mark inversion (AMI). The encoded sequence is obtained by representing the mark in the binary sequence alternately by positive and negative

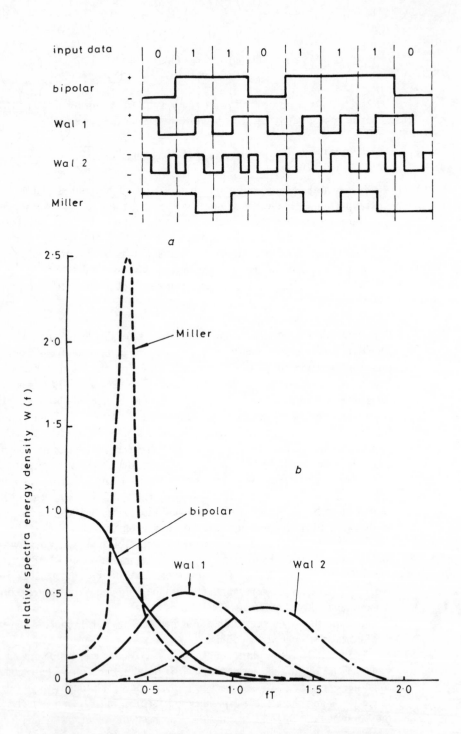

Fig. 6.12 – Binary codes based on Wal 1 and Wal 2. (a) Transmitted waveform. (b) Line spectra.

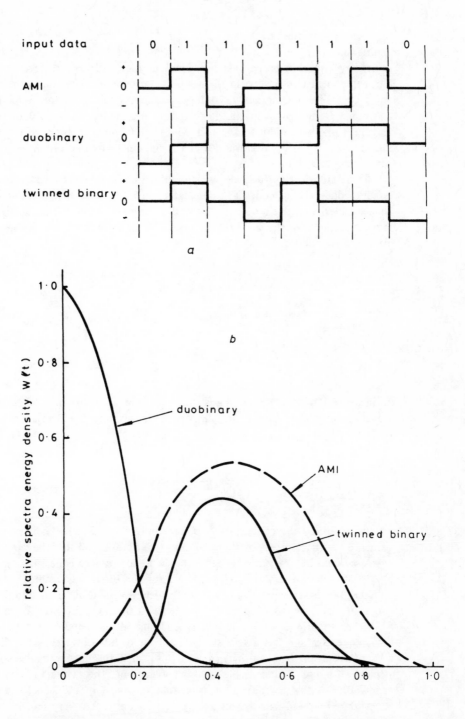

Fig. 6.13 — AMI and linear pseudo-ternary codes. (a) Transmitted waveforms. (b) Line spectra.

impulses, whilst the spaces are represented by no signal. The code structure and its line signal power spectrum are illustrated in Fig. 6.13. The AMI code has some very attractive properties. The line signal power density spectrum has no d.c. component and a very small l.f. spectrum content. The code disparity is equal to 1. The coding and decoding circuitry requirements are quite simple and some degree of error monitoring can be achieved by simply observing violations of the AMI rule. The bandwidth requirement is equal to the transmission rate. It has the disadvantage that it has a poor timing content associated with long runs of binary zeros.

There are some useful codes which fall within the subclass of linear pseudo-ternary codes. These codes are designated linear because the pseudo-ternary code is linearly derived directly from the binary message. A linear pseudo-ternary code is actually a particular case of a binary code in which the signal element $S_0(t)$ has been replaced by

$$S_0'(t) = S_1(t) \times S_0(t) ,$$

where $S_1(t)$ is the sequence of impulses

$$S_1(t) = \sum_{k=0}^{x} \alpha_k (t - kT) .$$

A linear pseudo-ternary encoding is thus equivalent to a filtering operation, the frequency response of the equivalent filter being given by the Fourier transform of $S_1(t)$

$$S_1(\omega) = \sum_{k=0}^{K} \alpha_k e^{-jkT}$$

In order for the coded signal to have only three possible values for any sequence it is necessary that there be only two $\alpha_k \neq 0$ and that they are either equal or opposite. We thus have two basic pseudo-ternary codes; the twinned binary, in which $\alpha_0 = -\frac{1}{2}$ and $\alpha_1 = +\frac{1}{2}$ (or the other way round), and the duobinary code, in which $\alpha_0 = \alpha_1 = +\frac{1}{2}$. The AMI and linear pseudo-ternary codes are illustrated in Fig. 6.13, together with the power spectra for random data using these codes. The duobinary code has all its energy concentrated at low frequencies and has a very strong d.c. component. However, since the significant spectrum bandwidth is equal to only half the data rate, the code is attractive for use in limited bandwidth applications. The twinned binary code bandwidth is equal to the transmission rate and most of the signal energy is concentrated around half the bit rate. The spectrum is thus very similar to that of AMI code. The code is easily generated and has an error-detecting capability since it obeys the AMI rule. The code is, however, sensitive to error in the decoding operation due to error propagation in the decoding circuit.

There are two classes of non-linear ternary codes which have application in the field of data transmission, namely alphabetic and non-alphabetic codes. In the alphabetic codes, n binary digits are taken together giving a signal element which can be regarded as a selection from an alphabet of 2^n possible characters. The character is then encoded into m ternary digits where $3^m > 2^n$. Such codes are normally described as 'nBmT codes'. The simplest of these codes is given by $n = m = 2$ and is generally known as 'pair selected ternary' (PST). The message signal is grouped in 2-bit words which are then coded in ternary is given in Table 6.1.

Table 6.1 — PST code translation

Binary word	Ternary word		Word digital sum
	Mode A	Mode B	
00	− +	− +	0
01	0 +	0 −	±1
10	+ 0	− 0	±1
11	+ −	+ −	0

It can be seen there is no change in the rate of transmission. As each word is selected, the appropriate Word Digital Sum is added to a running sum total. The mode of the next ternary transmission is determined by the polarity of this running total. If the sum is negative, mode A is selected and if positive, mode B is selected. Where the total is equal to zero, the mode remains unchanged. Under normal conditions the running total will thus never vary by more than 1 from the zero value and will alternate between positive and negative polarity when taking a value other than zero. The maximum range of sum variation is often referred to as the code disparity. Thus the PST code has a disparity of 2. The d.c. component in the transmitted signal is effectively reduced by the alternating mode, which is equivalent to the alternating polarity of bipolar coding. Timing content is assured by the translation of pairs of zeros into pulses. The mode alternation also provides some error monitoring capability. The average power spectrum is given in Fig. 6.14(a). The drawbacks of the code are that it has high l.f. components and the transmission power is about 1.5 times that of AMI for similar performance.

The ternary codes discussed so far are pseudo-ternary inasmuch as each ternary symbol only has a binary significance. A true ternary code is capable of conveying $\log_2 3$ bits of information per symbol rather than the binary rate of one bit per symbol. The pseudo-ternary codes are thus only 63 per cent efficient

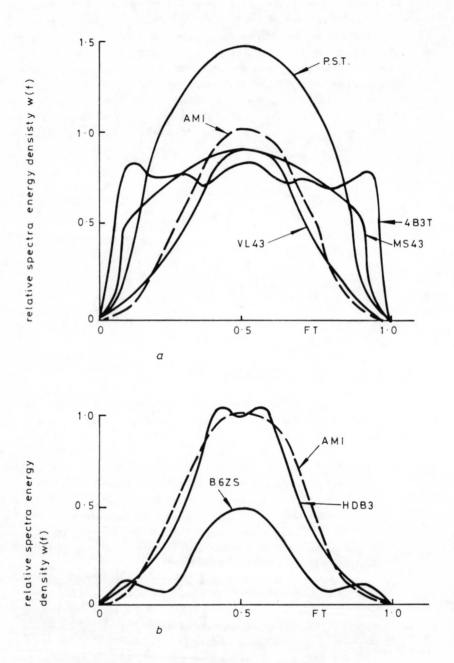

Fig. 6.14 – Power spectra for ternary coded random data. (a) Alphabetic ternary codes. (b) Non-alphabetic ternary codes.

in transmission capability, although the redundancy may give some error-detecting capability to the code. For example, in AMI, violations of the AMI rule would indicate some error in transmission had occurred.

The efficiency of transmission can be improved by the use of alphabetic codes where $m < n$. A widely used code of this class is that known as '4B3T'. The original binary data stream is divided into words of four bits, each word being encoded into three ternary digits as shown in Table 6.2.

Table 6.2 – 4B3T code translation

Binary word	Ternary code		Word digital sum
	Mode A	Mode B	
0000	+ 0 −	+ 0 −	0
0001	− + 0	− + 0	0
0010	0 − +	0 − +	0
0011	+ − 0̂	+ − 0	0
0100	+ + 0	− − 0	±2
0101	0 + +	0 − −	±2
0110	+ 0 +	− 0 −	±2
0111	+ + +	− − −	±3
1000	+ + −	− − +	±1
1001	− + +	+ − −	±1
1010	+ − +	− + −	±1
1011	+ 0 0	− 0 0	±1
1100	0 + 0	0 − 0	±1
1101	0 0 +	0 0 −	±1
1110	0 + −	0 + −	0
1111	− 0 +	− 0 +	0

The mode alternation is governed by the same rules as used for the PST code. The code gives a possibility of runs of similar ternary digits up to a maximum of 6. Computing the running sum on a digit by digit basis, the code disparity is 8. However, if the running sum is only observed at the end of each of the ternary word groups, then the maximum sum variation is restricted to 6. The average power spectrum of the 4B3T code with random data is given in Fig. 6.14(a). The power is fairly evenly distributed throughout the spectral band but there is a significantly large component at the low frequency end of the spectrum. Some attempts to overcome this large l.f. component have been made by the introduction of modified 4B3T-type codes. Two of these are the MS-43 code and the VL-43 code. Both these codes have a more complex arrangement for mode alternation than the basic 4B3T code described above. The average power spectra for the two codes are given in Fig. 6.14(a).

In the non-alphabetic codes, long runs of zeros which may occur in conventional AMI coding are broken up by the substitution of pulses or groups of pulses which violate the AMI alternating pulse polarity rule. There are a number of ways this may be carried out; the most widely used method being the code known as HDB3 ('high density binary' with maximum of three consecutive zeros). HDB3 is a modification of AMI where, if more than three consecutive zeros occur, a 'violation' pulse is substituted for the fourth zero. The first violation pulse is selected so that it indeed violates the alternate mark inversion rule and is thus readily identifiable as a zero rather than a mark. Subsequent violation pulses simply alternate in polarity from the first violation pulse to maintain low code disparity. This means, however, that some violation pulse substitutions are not identifiable from genuine marks. To avoid this confusion, any violation pulse which is not of opposite polarity to the preceding mark pulse is forced into violation by the insertion of a 'parity' pulse substituted for the first zero immediately following the genuine mark. The next mark following a parity pulse is then made of opposite polarity to the parity pulse, irrespective of the polarity of the previous genuine mark. In this way, unique decoding is possible. The procedure is probably more easily understood by considering a specific example. Consider the bit sequence given in Fig. 6.15. The first violation pulse V_1 occurs in place of the fourth zero following the first mark pulse. It is of opposite polarity to the initial mark. The second violation pulse V_2 occurs in place of the fourth zero in the next run of zeros and is of opposite polarity to the previous violation pulse. However, it does not violate the inversion rule with respect to the preceding mark pulse. Parity pulse P_1 is therefore inserted in place of the first zero in the run of zeros. The polarity of this pulse is chosen to cause violation by V_2. Obviously, on receipt, P_1 will first be considered as a mark. However, on detection of the violation pulse V_2, it will be clear that P_1 must have been a parity pulse and should therefore be interpreted as a zero. The next mark following the parity pulse is of opposite polarity to the parity pulse, the parity pulses being regarded as marks as far as the AMI rule is concerned, even though they in fact represent a zero value in the data sequence. Thus, following the

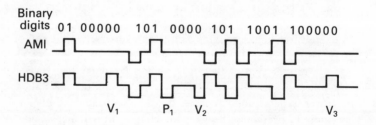

Fig. 6.15 — HDB3 code.

parity pulse P_1 in Fig. 6.15, the marks are of opposite polarity to what they would have been had conventional AMI code have been used. Finally, violation pulse V_3 occurs in place of the fourth zero in the third sequence of zeros. This pulse is of opposite polarity to V_2 and violates the alternate mark inversion rule with respect to the preceding mark pulse. No parity pulse is therefore necessary.

The average power spectrum for HDB3 is given in Fig. 6.14(b), together with that of an alternative code known as B6ZS ('binary with six zeros substitution'). The B6ZS code is similar to HDB3, where any run of six consecutive zeros is replaced by a code group of pulses which can be identified at the receiver because some of the substituted pulses violate the basic AMI rule.

Codes using more than three pulse amplitude levels are possible. For instance, quarternary (four-level) codes can be simply derived by taking pairs of bits from the binary input sequence and converting them into pulses with amplitudes -3, -1, $+1$ and $+3$ corresponding to the binary pairs 01, 00, 10 and 11 respectively. This procedure can be extended by taking n bits at a time and converting them into an m-ary signal where $m = 2^n$.

Another technique using multilevel signalling is the generalised partial response coding. The linear pseudo-ternary codes described earlier are the simplest forms of partial response coding. Higher order partial response codes make use of an increasing number of amplitude levels to define the signal. The main disadvantage of the multilevel codes is that the higher number of levels makes them vulnerable to interference from external sources such as crosstalk, impulsive noise and radio pick-up.

MODULATION OF DATA SIGNALS

Generally basic data signals do not exist in a form that is suitable for direct transmission over telephone connections. For instance, the normal telephone connection is not a low-pass circuit since there are exchange transmission bridges and other forms of a.c. coupling associated with almost any telephone circuit. Since a typical data signal has significant frequency components near to d.c., it is therefore impossible to carry such signals directly over the telephone network. To combat this problem, modulation – demodulation equipment is used. Such an equipment is normally referred to as a 'modem'. The modem converts data signals to freqeuncies in the voice-band which are suitable for transmission over the telephone network and on reception recovers the transmitted data sequence from the received signal.

For low-speed data applications frequency-shift-keying (FSK) is a reliable, easily implemented modulation scheme. In this scheme the binary data 0s and 1s are represented by two frequencies f_1 and f_2, respectively. FSK is really a special case of frequency modulation, where the modulating signal is a binary data sequence. It therefore produces a large number of significant side-bands, and thus does not make optimum use of the available bandwidth. It is not used, therefore, for high-speed data applications. The alternatives to frequency-shift-keying are amplitude modulation and phase-shift-keying (PSK).

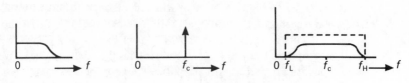

Base-band signal × Cosine carrier = Bandpass signal

(a)

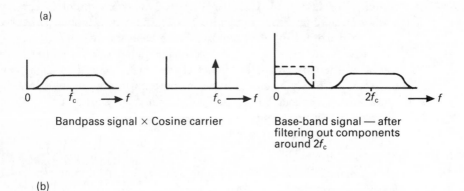

Bandpass signal × Cosine carrier   Base-band signal — after filtering out components around $2f_c$

(b)

Fig. 6.16 – DSBSC modulation and demodulation. (a) DSBSC modulation. (b) DSBSC synchronous demodulation.

Amplitude modulated signals are generated by multiplying the base-band signal by a cosine carrier of frequency $f_c = (f_H + f_L)/2$, the arithmetic mean of the edge frequencies of the available transmission bandwidth. We thus produce a double side-band suppressed carrier signal (DSBSC) as shown in Fig. 6.16(a). Provided the bandwidth of the bandpass channel is at least twice the bandwidth of the low-pass signal, i.e. $f_H - f_L > 2W$, the modulated signal will go through the channel unaffected by its band-limiting. The original base-band signal can then be recovered by multiplying the bandpass signal with a cosine carrier and low-pass filtering the resultant product waveform as illustrated in Fig. 6.16(b).

Since amplitude modulation is a linear process, the band pass channel, sandwiched between a suitable modulator and demodulator, behaves externally just like a low-pass channel. The equivalence is shown in Fig. 6.17. This enables any techniques available for low-pass channels to be applied directly to bandpass channels. Thus channel shaping and equalisation can be carried out just as though the transmission were through a base-band channel. Amplitude modulation may be carried out at two levels by direct modulation with the binary data stream, or multilevel modulation can be used where digits are taken from the data stream n at a time and then transmitted as a 2^n amplitude-modulated signal.

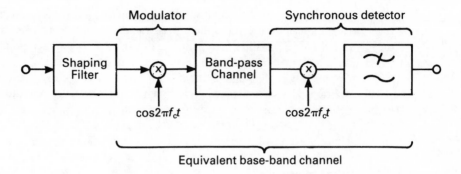

Fig. 6.17 – Equivalence of bandpass and base-band channels.

Since amplitude modulation produces two side-bands containing identical information, bandwidth utilisation may be improved by the use of single-band (SSB) techniques. Since the modulating signal may contain significant low frequency components, it is usual to use vestigial side-band (VSB) amplitude modulation rather then the suppression of the complete side-band as is usual with SSB.

Phase-shift-keying involves taking n bits at a time from the data sequence and encoding into 2^n phase-shifts. Signal space diagrams for 8 level amplitude modulated and 8 level PSK signals are given in Fig. 6.18. The circles of radius E_e define the maximum permissible perturbation from the ideal signal before an error in transmission may occur. It is obvious from the diagram that PSK has an advantage in terms of error threshold over the equivalent amplitude modulation

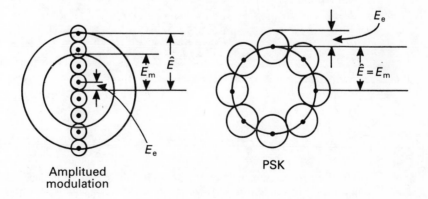

Fig. 6.18 – Signal space diagram, amplitude modulation and PSK (8-state).

scheme. Since, in multilevel amplitude modulation and PSK, digits are taken from the data stream n at a time, the line signal modulation rate (baud rate) is only $1/n$ of the data signalling rate. Thus the line bandwidth required is reduced by a factor n compared to that required for binary modulation.

By a combination of amplitude modulation and PSK, it is possible to choose signal conditions that give a much greater margin of tolerance to perturbation before the onset of errors than for either amplitude or phase modulation alone. Examples for 8-, 16- and 32-state signalling are given in Fig. 6.19. Comparing the error thresholds for these combined 'hybrid' modulation schemes with those for the equivalent phase and amplitude modulation schemes, we can clearly see the advantage to be gained in terms of noise immunity using hybrid modulation. The advantage is, however, at the cost of more complex modulation and de-modulation equipment.

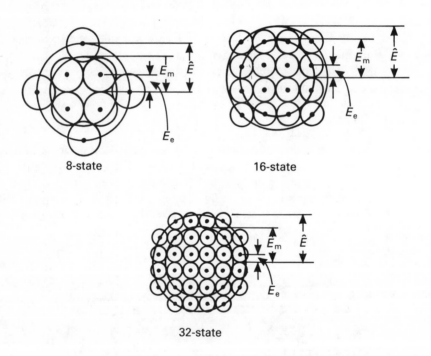

8-state   16-state

32-state

Fig. 6.19 – Signal space diagram, 'M' state hybrid modulation.

It is often convenient to consider a hybrid modulation scheme as a pair of independent DSBSC signals having the same carrier frequency but with the carriers 90° out of phase with each other. Thus a bandpass channel can be sand-wiched between a pair of synchronous modulator/demodulator systems as shown in Fig. 6.20. Such an arrangement is often referred to as 'quadrature-

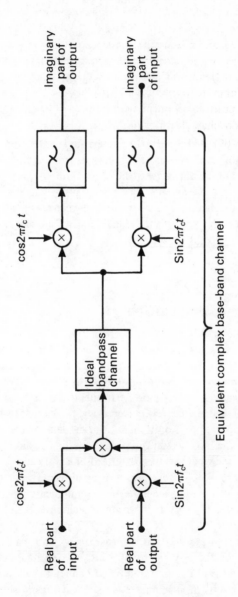

Fig. 6.20 — Quadrature amplitude modulation system (QAM).

amplitude modulation' (QAM). The result is equivalent to a base-band channel that is capable of handling simultaneously a pair of waveforms. This can conveniently be regarded as a base-band channel which can handle complex waveforms; a complex waveform is simply a pair of real waveforms treated in an appropriate way. Thus the 16-state hybrid signal of Fig. 6.19, which provides 4 bits per symbol, can be regarded as two quadrature 4-level (2-bit) signals with pulse heights ±1 and ±3. Similarly, the 32-state hybrid signal may be regarded as two quadrature 6 level signals with one unused signal point in each quadrant. The equivalent base-band channel will generally have an impulse response that is complex, that is, a purely real input applied to the equivalent base-band channel will produce an output with non-zero imaginary part. The exception is when the gain characteristic of the bandpass channel is symmetric about the carrier frequency and the phase characteristic is antisymmetric about it. There is thus some cross-modulation of the data carried on the phase and quadrature channels. To minimise this effect, the channel impulse response should be maintained as nearly real as possible. The main cause of cross-modulation is an error in the phase of the synchronous demodulating carrier signal. A little later we will show how a complex transversal equaliser can be used to eliminate unwanted signals caused by cross-modulation.

DATA TRANSMISSION SERVICES

In the early 1960s British Telecom (then the GPO), together with other telecommunication carriers throughout the world, introduced data transmission services which, in the UK, came under the generic title of Datel Services.

The first modems to be introduced into service were FSK modems. FSK was chosen because it gave a good performance, was reasonably simple to implement, provided stable operation over a wide variety of channels and permits asynchronous operation. Additionally, the modem performance is not affected in any way by the data sequence because the line signal is always present with constant amplitude whatever the pattern of binary 1s and 0s.

Two standard FSK systems are in general use. The first was designed to enable keyboards and similar machines to communicate over the public switched telephone network (PSTN) and private speech circuits at signalling rates of up to 200 bits/s in a full duplex mode. CCITT recommendation V21 was established to cover this requirement internationally and made provision for a possible extension of the rate up to 300 bits/s in each direction. The frequency allocation chosen is given in Table 6.3. Channel 1 is used for the transmission of the caller's data while channel 2 is used for transmission in the other direction. Asynchronous or synchronous working between data equipments is possible when using these modems. The UK version of this modem is known as Datel Modem No. 2 and in the USA the Bell System Dataset No. 103 provides similar facilities.

Table **6.3** — Characteristic frequencies for V21 modems operating at 200 bits/s over PSTN circuits.

Channel	Nominal mean frequency (Hz)	Binary symbol 1 frequency (Hz)	Binary symbol 0 frequency (Hz)
1	1080	980	1180
2	1750	1650	1850

The second standard was introduced to provide a higher-speed link, the main use of which has been to enable visual display units to be serviced by a host computer with greater rapidity than was possible by the previously described modem. A low-speed return channel, which could allow control signals or telegraph-type keyboard signals to be transmitted, was provided as an option if required. The modem again operated on either the public switched telephone network or private speech circuits, this time at data signalling rates of up to 1200 bits/s. However, with a small proportion of the poorer quality switched network connections, difficulty was experienced in obtaining 1200 bits/s and so a fall-back facility was provided to enable operation at 600 bits/s to be assured over the PSTN. The return channel operates at modulation rates of up to 75 bits/s. The frequency allocation for this modem is given in Table 6.4. The international standard is established in CCITT recommendation V23. In the UK, Datel Modem No. 20, which supercedes Modem No. 1, conforms to this recommendation and in the USA Bell System Data Set No. 202 fulfils the same requirement.

Table **6.4** — Characteristic frequencies for V23 modems operating at 600/1200 bits/s.

Mode	Nominal mean frequency (Hz)	Binary symbol 1 frequency (Hz)	Binary symbol 0 frequency (Hz)
1 (up to 600 bits/s)	1500	1300	1700
2 (up to 1200 bits/s)	1700	1300	2100
Optional return channel (75 bits/s)	420	390	450

For data signalling rates in excess of 1200 bits/s, techniques which make more efficient use of the available bandwidth have to be employed. Examples of modulation systems which require less line bandwidth for a given modulation rate are single-side-band (SSB) amplitude modulation, vestigial-side-band (VSB) amplitude modulation, phase shift keying (PSK) and quadrature amplitude modulation (QAM).

With all of these more complex modulation techniques it is generally necessary to equalise the line or channel characteristics and, especially for the higher data rates, this often has to take the form of automatic adaptive equalisation, which is able to adjust to any changes in the channel characteristics as they occur.

The first requirement for a data rate greater than 1200 bits/s was met by a modem designed to operate at 2400 bits/s. Originally this modem, to CCITT recommendation V26, was intended for operation over four-wire private speech band circuits. However, a variant (V26 bis) was subsequently introduced for use over the public switched telephone network. This variant was provided with a fall-back facility to 1200 bits/s where the line characteristics were unsuitable for 2400 bits/s operation. The modem operated at a modulation rate of 1200 bauds, the data being taken in pairs of bits and being conveyed as one of four possible phase-shifts of a 1800 Hz carrier. The four-phase signal is differentially encoded to avoid the need to transmit a reference phase signal for the purpose of demodulation. The information is thus decoded at the receiver by comparing the phase of the carrier preceding the instant of modulation with that following it. The significance of phase changes is given in Table 6.5. The two alternatives are available for the leased line modem (V26) but the switched network modem (V26 bis) is only available using alternative B. The advantage of alternative B is that there is always a phase-change between adjacent modulation epochs.

Table 6.5 – Line-signal phase changes for V26 differential four-phase modem.

Pair of data signal elements (dibit)	Phase change (degrees)	
	Alternative A	Alternative B
00	0	+ 45
01	+ 90	+135
11	+180	+225
10	+270	+315

The nature of the four-phase differential modulation techniques used in this modem, in which pairs of data signal-elements are identified by phase changes of the carrier signal, requires that the transmission by synchronous, i.e. that both the modulator and demodulator, together with the data terminal equipment, must be controlled by timing signals in such a way that the individual data-signal elements can be correctly identified in the demodulation process. Within the UK the Datel Modem No. 12 and in the USA the Bell Data Set No. 201 conform with the recommendation.

Following the successful operation of 2400 bits/s modems the next requirement is for 4800 bits/s operation. This is achieved by the use of differential eight-phase modulation on a 1800 Hz carrier at a modulation rate of 1600 bauds. Equalisation is necessary in order to obtain satisfactory operation at this data rate. Three alternatives have been recommended by CCITT as follows:

V27   4800bits/s modem with manually set equaliser for use on high quality leased voice-band circuits.

V27 bis   4800 bits/s modem with automatic adaptive equaliser and fall-back facility to 2400 bits/s operation in accordance with recommendation V26A for use on general quality leased voice-band circuits.

V27 ter   4800 bits/s modem with automatic adaptive equaliser and fall-back facility to 2400 bits/s operation in accordance with recommendation V26A for operation over the public switched telephone network.

The significance of the phase changes for each of these options is given in Table 6.6.

Table 6.6 — Line signal phase changes for differential eight-phase modems.

Tribit value	Phase change (degrees)
001	+0
000	+45
010	+90
011	+135
111	+180;
110	+225
100	+270
101	+315

The inevitable demand for even higher rates then led to a further recommendation for operation at 9600 bits/s over selected high-quality voice-band lines. This recommendation, CCITT recommendation V29, specifies a combined phase and amplitude modulation system operating at a modulation rate of 2400 bauds. The data is taken in groups of four consecutive data bits (quadbits). The first bit (Q1) in time of each quadbit is used to determine the signal element amplitude to be transmitted. The second (Q2), third (Q3) and fourth (Q4) bits are encoded as a phase change relative to the phase of the immediately preceding element as shown in Table 6.6 for the V27 4800 bits/s modem. The relative amplitude of the transmitted signal element is determined by the first bit (Q1) of the quadbit and the absolute phase of the signal element. The absolute phase is initially established by means of a synchronising signal transmitted prior to the transmission of actual data. The relative signal element amplitudes are given in Table 6.7.

Table 6.7 — Relative amplitude of signal elements, V29 modem.

Absolute phase	Q1	Relative signal
0, 90, 180, 270	0	3
	1	5
45, 135, 225, 315	0	$\sqrt{2}$
	1	$3\sqrt{2}$

The 16-point signal space constellation associated with this signalling arrangement is illustrated in Fig. 6.21. Fall-back rates of 7200 bits/s and 4800 bits/s are specified. For 7200 bits/s operation, the outer ring of signal states is not used. To achieve this, the data is taken in tribits which are used to determine bits Q2 to Q4 of the modulator quadbit, whilst Q1 is held at the 'zero' condition. For 4800 bits/s operation, the data is taken in dibits and specifies only the four states of relative amplitude 3 in phase and quadrature with the absolute carrier. The carrier frequency at all data rates is 1700 Hz. A fully automatic adaptive equaliser, as described in the next section, is required with this modem. Modems operating at rates of 9600 bits/s are often used as part of a multiplexed digital transmission system, thus enabling a high transmission efficiency to be obtained.

EQUALISATION

Imperfection in the transmission channel characteristics will inevitably lead to departures from the ideal channel impulse response we have endeavoured to obtain for our equivalent base-band channel by careful design of the channel-

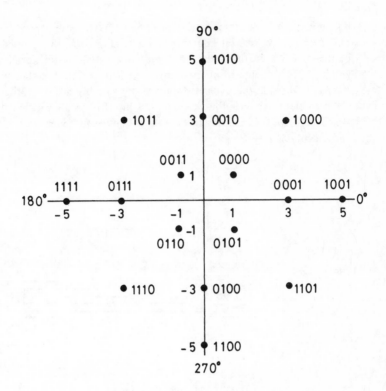

Fig. 6.21 – Signal space constellation for 9600 bit/s V29 modem.

shaping filters. Instead of passing through zero at intervals of time T, the impulse response takes on values as shown in Fig. 6.22. This gives rise to inter-symbol interference if consecutive data pulses are transmitted at intervals of time T.

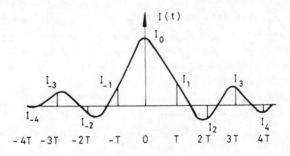

Fig. 6.22 – Impulse response having inter-symbol interference components.

This inter-symbol interference can be reduced by the use of a transversal equaliser. The basic transversal equaliser consists of a delay-line tapped at intervals of time T. Each tap has associated with it a variable gain device as shown in Fig. 6.23. The outputs from these multipliers are added together in a summing network to provide the equalised output. By suitably adjusting the gain setting it is possible to reduce the inter-symbol interference. Enough delay-line taps have to be provided in order to be able to reduce the inter-symbol interference to an acceptable level.

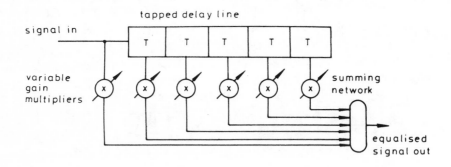

Fig. 6.23 — Basic transversal equaliser.

Frequently the transmission circuit to be used, and hence the channel characteristics are not known until the sender is ready to transmit data over the network. This is the case, for instance, for dial-up circuits through the public switched telephone network. This means that equalisation has to be carried out each time a circuit is acquired for data transmission purposes. Sometimes the circuit characteristics are time-varying at a rate which allows significant changes to take place in the period of time the circuit is occupied for data transmission purposes. Under these conditions it is necessary to make the equalisation process adaptive. To automatically preset the equaliser, a test pattern is transmitted and the prior knowledge of the pattern at the receiver is used to compute the impulse response from the received signal. From the impulse response so obtained it is then possible either to calculate the tap coefficients directly or to use an iterative technique to successively increment the coefficients until an optimum setting is obtained. This strategy is illustrated in Fig. 6.24. This mode of operation gives fast initial setting-up but, if the channel characteristics are changing, it becomes necessary to retransmit the test pattern at intervals to allow for resetting of the coefficients. Fig. 6.25 shows an alternative mode of operation generally referred to as adaptive equalisation. Instead of using a test pattern the method is based on the assessment of the error between the actual received

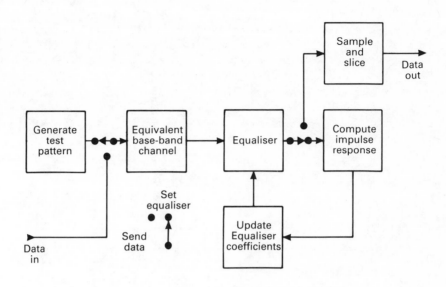

Fig. 6.24 – Automatic pre-set equaliser strategy.

signal and the receiver estimate of the transmitted symbol. The error signal obtained is correlated with the received data signal to obtain estimates of co-efficient setting errors. The coefficients are updated in accordance with these estimates to minimise the magnitude of the error signal. This strategy has the advantage that the setting-up procedure is a continually adapting feedback controlled process. However, if the unequalised signal is so badly distorted that a large proportion of erroneous decisions are made, the equaliser may not initially converge and equalisation will not be achieved. Under such conditions it is

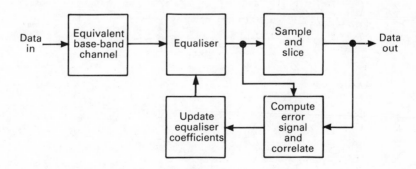

Fig. 6.25 – Adaptive equaliser strategy.

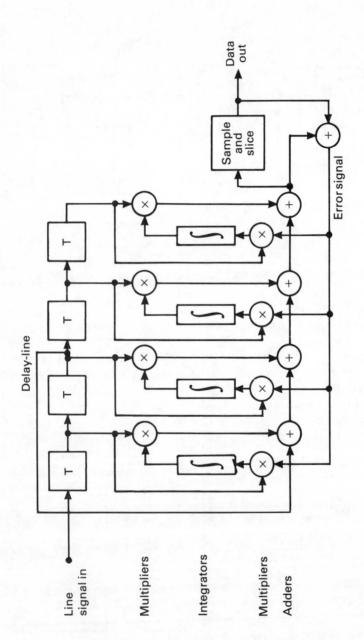

Fig. 6.26 — Basic hardware for adaptive equaliser.

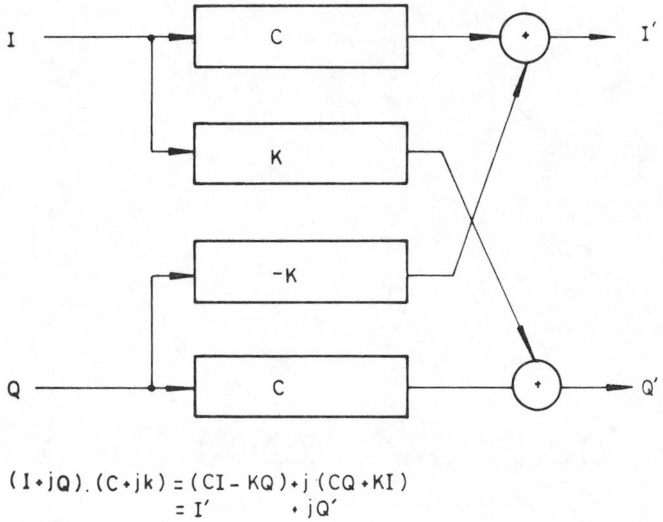

$$(I + jQ) . (C + jk) = (CI - KQ) + j (CQ + KI)$$
$$= I' \qquad + jQ'$$

Fig. 6.27 – QAM equaliser.

possible to start up in the preset mode, using an initial test pattern, and then to change to the adaptive mode when a reasonable degree of equalisation has been achieved. The basic hardware for the adaptive equaliser is given in Fig. 6.26. It can be shown that, provided the receiver estimate of the transmitted data is predominantly correct, then the equaliser converges to minimise the mean-square error in the received signal, whether the error arises from inter-symbol interference or from additive noise.

Equalisation of modulated data signals is usually carried out after demodulation back into base-band. For quadrature amplitude and phase modulated signals, the demodulated signal can be resolved into two quadrature components. Each component can then be equalised separately. There will also be some cross-modulation components which can be dealt with in a similar fashion to the interfering components in the main signal. A schematic diagram is given in Fig. 6.27. Since both signal components are operated on by similar channel characteristics, it is often convenient to consider the two input signals as the real and imaginery parts of a complex signal and, likewise, the main and cross-modulation coefficients of the equaliser as the real and imaginary parts of an array of complex equaliser coefficients. Thus:

$$(I + jQ) . (C + jK) = I' + jQ', \text{ as shown in Fig. 6.27.}$$

7

Pulse code modulation

If we transmit an analogue signal, such as a speech signal, over a channel with a finite transmission loss and subject to noise impairment, then, as the length of the transmission path is increased, the received signal-to-noise ratio decreases. Ultimately the signal may be completely masked by the additive noise.

Shannon's equation would suggest that if we could increase the bandwidth of our signal we could convey the same information with a lower signal-to-noise ratio. If we were to sample our original signal and represent the sample amplitudes digitally, using binary digits, we should indeed have increased our signal bandwidth. However, we now only have to be able to recognise the presence, or polarity, of pulses in a pulse train. This we can do with reasonable accuracy at levels of noise which would certainly have completely masked our original signal. What is more, instead of intermediate amplifiers, which increase the noise as well as the signal, we can install regenerators along the line. These regenerate the pulse train entirely free from noise. We thus have a technique which will enable us to transmit signals over channels which would otherwise be useless because of the noise impairment to the signal.

SAMPLING

If we are going to be able to reconstitute our original signal from our sample values without distortion, then the Sampling theorem tells us that we have to sample our signal at a rate at least twice as high as the highest frequency com-

ponent present in the signal. If we do not do this we shall create distortion due
to aliasing. A normal speech signal contains frequency components as high as
10 kHz but adequate intelligibility is maintained if the upper frequency limit is
around 3.4 kHz. Normal telephone quality speech is limited to this frequency
range. Providing, therefore, we filter our speech signal before sampling with an
anti-aliasing filter having a cut-off around 3.4 kHz, we can safely sample at a
rate of 8 kHz without introducing further distortion. Sampling of television
video signals is normally carried out at 10.2 MHz.

QUANTISATION

Our signal samples at this stage can take on an infinite range of amplitude
levels and are thus just as susceptible to noise as the original signal. We can,
however, at this point quantise the signal amplitude to the nearest of a range of
discrete amplitude levels as shown in Fig. 7.1. Obviously the quantisation
operation introduces a distortion into the signal which we refer to as 'quantisation noise'. The magnitude of this noise is a function of the number of quantisation levels used. For telephone speech 256 linearly spaced quantisation levels
would give a quality such that the quantisation noise would be hardly per-
ceptible above the other background noises normally associated with the telephone network. For television of the quality of normal broadcast colour television, 512 linearly spaced quantisation levels are required, while 64 levels gives
only fairly good television performance.

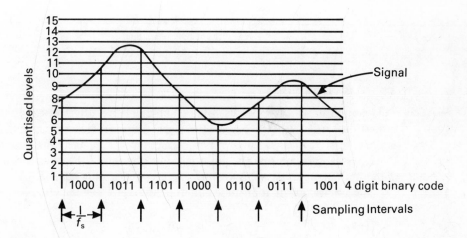

Fig. 7.1 – Quantisation of analogue signal.

REPRESENTATION BY BINARY SIGNALS

In order to take the greatest advantage of the noise immunity of digital signals, it is normal to transmit the quantisation level in binary code form. This is illustrated in Fig. 7.2. The number of binary digits required to represent a single sample is equal to $\log_2 L$, where L is the number of quantisation levels. The overall bit transmission rate is therefore $\log_2 L \times$ sampling rate. In practice, additional 'framing' bits are required in order to determine which bits in the continuous stream are related together to indicate a single sample. In early PCM systems, 128 amplitude levels were considered adequate and thus 7 bits per sample were required to define the signal amplitude. An extra supernumerary bit was then added to each sample to give an octet of 8 bits per sample. Thus the overall bit rate per speech channel was $8 \times 8 = 64$ kbits/s. In later systems the number of bits used to represent the signal amplitude was increased to 8, giving 256 levels. However, the basic octet structure was maintained and the supernumerary bits were included in separate octets designated especially for signalling and synchronisation. The details are given later when PCM multiplexing is discussed in more detail.

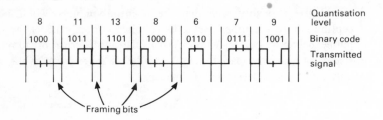

Fig. 7.2 – Binary encoding.

QUANTISATION NOISE

Quantisation noise arises because of the approximation effect of choosing the nearest quanatisation level to the actual signal value. Providing the quantisation is linear, that is the step interval is the same throughout the signal amplitude range, we can assume that the probability distribution of the error is constant within the range plus or minus half the step size. This is shown in Fig. 7.3.

The average quantisation noise output power is given by the variance

$$\sigma^2 = \int_{-\infty}^{\infty} (e - \mu)^2 p(e)\, de \ ,$$

where μ = mean. This will be zero for quantisation noise. Given that the quantisation step size is δV,

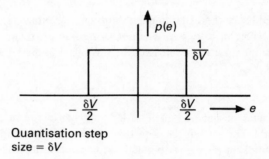

Quantisation step
size = δV

Fig. 7.3 — Probability distribution of error due to linear quantisation.

$$p(e) = \frac{1}{\delta V} , \quad -\frac{\delta V}{2} \leqslant e \leqslant \frac{\delta V}{2} .$$

$$\sigma^2 = \int_{-\frac{\delta V}{2}}^{\frac{\delta V}{2}} (e-0)^2 \frac{1}{\delta V} \, de$$

$$\frac{1}{\delta V} \int_{-\frac{\delta V}{2}}^{\frac{\delta V}{2}} e^2 \, de$$

$$\frac{1}{\delta V} \left[\frac{e^3}{3} \right]_{-\delta V/2}^{\delta V/2}$$

$$\frac{(\delta V)^2}{12} .$$

To give an indication of the magnitude of the quantisation noise power in relation to a given power, consider a sinusoidal signal of peak-to-peak amplitude $2V$ quantised into L levels. For linear quantisation, the step size $\delta V = (2V)/L$ and therefore the rms signal-to-quantisation noise ratio is

$$10 \log_{10} \left(\left(\frac{V}{\sqrt{2}} \right)^2 \Big/ \frac{(\delta V)^2}{12} \right) \text{dB}$$

$$= 10 \log_{10} \left(\frac{V^2}{2} \Big/ \frac{4V^2}{12L^2} \right) \text{dB}$$

$$= 10 \log_{10} \left(\frac{V^2}{2} \times \frac{12L^2}{4V^2} \right) \text{dB}$$

$$= 10 \log_{10} (1.5L^2) \text{dB}$$

$$= 20 \log_{10} (1.225L) \text{dB}$$

and hence the signal-to-quantisation noise voltage ratio $= 1.225L$.

But $n = \log_2 L$, where n is the number of binary digits used to represent the quantisation level. Thus the rms signal-to-quantisation noise ratio is given by

$$1.8 + 6n \text{ dB}$$

Values of signal-to-quantisation noise ratio for various code sizes are given in Table 7.1. It will be seen that every additional code bit gives an improvement of 6dB in the signal-to-quantisation noise ratio. These figures, of course, assume that the signal occupies the full dynamic range of the system and that over the duration of each cycle, the signal is approximately sinusoidal.

Table 7.1 – Quantisation noise vs. binary code size.

Binary code size n (bits)	No. of quantising levels $= 2^n$	Signal-to-quantisation noise ratio	
		Voltage ratio	dB
2	4	4.9	13.8
3	8	9.8	19.8
4	16	19.6	25.8
5	32	39.2	31.8
6	64	78.3	32.8
7	128	157	43.8
8	256	314	49.8
9	512	627	55.8
10	1024	1251	61.8

COMPANDING

In practice speech signals vary very considerably in amplitude, depending on the word spoken and the loudness of the speaker concerned. The problem with linear quantisation is that the magnitude of the quantisation noise is absolute for a particular system and does not depend on signal amplitude. Comparatively, therefore, weak, low-level signals suffer worse from quantisation noise than the loud, stronger signals. To overcome this problem it is possible to quantise using non-linear step sizes which are considerably smaller for weak signals than for strong signals. The quantisation noise is then dependent on the signal amplitude. We can thus reduce the number of code bits required to give a certain 'fineness' of quantisation at low signal levels. Because the effect is to compress large signals for transmission and to expand them again on receipt, the process has become known as 'companding'.

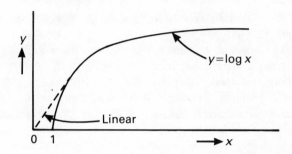

Fig. 7.4 – Function $y = \log x$.

Our knowledge and use of the decibel would suggest that a logarithmic law would be appropriate for companding. Unfortunately, as we can see from Fig. 7.4, the function $y = \log x$ does not pass through the origin. It is therefore necessary to substitute a linear portion to the curve for lower values of x. Most practical companding systems are based on a law suggested by K. W. Cattermole, namely

$$y = \frac{1 + \log_n Ax}{1 + \log_n A} \quad \text{for } \frac{1}{A} \leqslant x \leqslant 1,$$
$$\text{(logarithmic section)}$$

and
$$y = \frac{Ax}{1 + \log_n A} \quad \text{for } 0 \leqslant x \leqslant \frac{1}{A}$$
$$\text{(linear section)}.$$

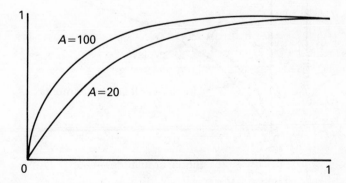

Fig. 7.5 – Cattermole's companding law.

A is the compression coefficient and the curve is continuous at $x = 1/A$. The law is illustrated in Fig. 7.5.

The practical implementation of this law would require non-linear signal processing followed by linear quantisation. The implementation of the non-linearity, together with the complementary non-linearity at the receiver, presents considerable problems. It is usual, therefore, to implement a piece-wise linear segmental approximation to the law. Two piece-wise linear laws are in

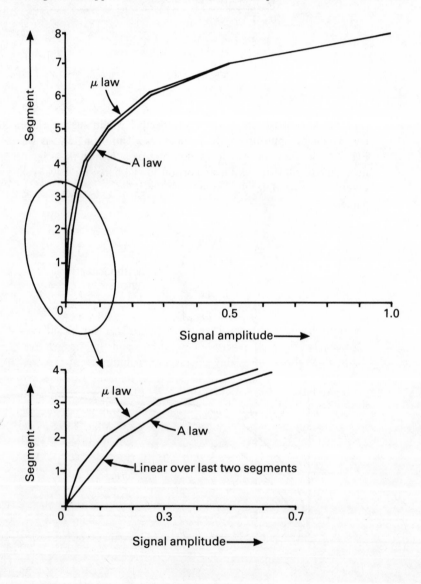

Fig. 7.6 – A law and μ law companding.

general use. That used in Europe is that known as CCITT companding law A. A law companding consists of eight linear segments for each polarity, the slope halving for each segment except the lowest two segments. The law is shown diagrammatically in Fig. 7.6. Three bits are required to define the segment. Within each segment there are 16 linear steps (8 steps for early systems). These 16 steps are defined by a further 4 bits. Finally, the polarity of the signal is defined by 1 bit, giving a total of 8 bits in all. There are thus 256 defined signal levels. Because the full curve has in all 13 linear segments, A law companding is sometimes referred to as 13 segment companding. The compression coefficient for A law companding is approximately equal to 87.6. In the USA a slightly different companding law known as μ law companding is widely used. In μ law companding the slope is halved over all eight segments as shown in Fig. 7.6. Otherwise, the procedure is very similar to A law. Because the full curve comprises 15 linear segments, μ law companding is sometimes referred to as 15 segment companding. It will be seen from Fig. 7.6 that there is hardly any significant difference between A law and μ law companding except at very low signal amplitudes.

The piece-wise linear companding rules can be readily implemented within the analogue-to-digital conversion process. As the signal sample passes up from one segment to another, all that is required is a doubling of the step size of the analogue-to-digital converter. This can be achieved by simply discarding the least significant bit of the converter output whenever a segment threshold is passed.

THE USE OF PCM IN THE TELEPHONE NETWORK

Pulse-code-modulation is now the preferred method of transmission for use in the junction and trunk transmission network. In the UK the plan is that the whole junction and trunk network should be converted to digital transmission within the next decade. There is at present no significant use of pcm in the local network, that is between subscriber and the local exchange. Because of this, PCM almost invariably operates on a multiplex basis. When PCM was first introduced into Britain, British Telecom (then the British Post Office) chose a 24-channel standard based on the earlier FDM group concept. 7-bit samples with an additional supernumerary bit per octet was used. Thus the actual digital transmission rate was $24 \times 64 \times 10^3 = 1.536$ Mbits/s. At a later date an international standard was issued by CEPT which considered 8 bits per sample was more appropriate for acceptable quality speech. This standard recommended 30 channels per cable pair, but the supernumerary bits are removed from the sample packets so that the basic channel rate was still left at 64 kbits/s. Instead, two extra channels were introduced to carry the framing and signalling information, so that the overall cable transmission rate was $(30 + 2) \times 64 \times 10^3 = 2.048$ Mbits/s. Both standards operating in the UK use A law companding. All channel equipment is gradually being replaced with 30-channel standard equipment. Besides the greater capacity, the removal of the signalling information from the channel gives added flexibility when used in conjunction with digital switching.

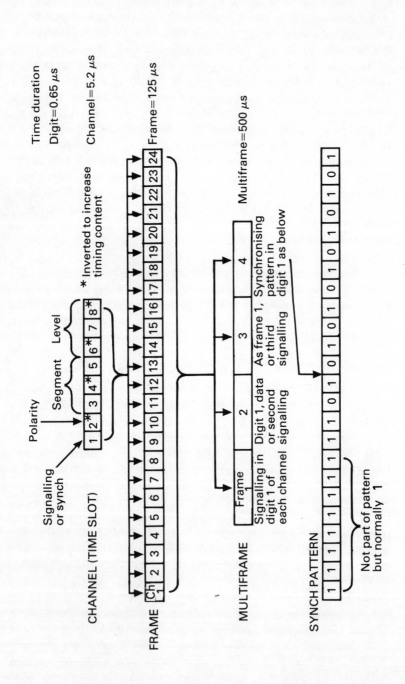

Fig. 7.7 – 24-channel PCM digit assignment.

Ternary line transmission codes are normally used for PCM. The 1.536 Mbits/s transmission for 24-channel operation normally uses AMI, or its derivative HDB3. However, if 4B3T is used for the 2.048 Mbits/s transmission required for 30-channel PCM, then the ternary symbol rate is 2.048 × ¾ = 1.536 Msymbols/s, the same line transmission rate as used for 24-channel PCM. The conversion from 24-channel to 30-channel operation can thus be carried out without the necessity to change the line repeater design and spacing. This feature of 4B3T encoding makes it an attractive proposition despite its somewhat complex decoding rules and the need for symbol segmentation.

The optimum repeater spacing for the transmission of PCM at a ternary symbol rate of 1.536 Msymbols/s over junction cable happens to be about 2000 yards, the same spacing as used for loading coils for conventional analogue speech transmission. It is thus possible to use the same manholes for the repeaters as were used for the loading coils when circuits are converted from analogue to digital transmission. The loaded cable pair carries only one telephone circuit, whereas the 30-channel PCM circuit carries one direction of transmission of 30 separate telephone conversations. A further cable pair is needed for the return channels for full-duplex operation. Transmission is limited to one direction of transmission per circuit pair because repeaters are intrinsically unidirectional devices.

DIGIT ASSIGNMENT

Details of the digit assignment used is 24-channel PCM equipment is given in Fig. 7.7. Because of pauses between sentences, words and even phonemes, the most frequently occurring sample amplitude on any telephone circuit is that representing silence. This gives rise to an idling pattern which would, with normal encoding procedures, consist mainly of zeros. Such a pattern will have few transitions from which to deduce timing information. The timing information content can be significantly increased by inverting every alternate bit in the basic octet, giving an idling pattern consisting substantially of signal reversals. Frame synchronisation is obtained by introducing a recognisable sequence of bits into the supernumerary digit 1 slots on each channel in every fourth frame. This gives a four-frame multi-frame data structure. In the other three frames, the supernumerary digits are used for channel signalling, the signalling occurring in the same time-slot as the related channel sample. Up to three different conditions can be signalled by using sequentially each of the frames 1 to 3 in each multi-frame.

The digit assignment normally used for 30-channel PCM systems is shown in Fig. 7.8. There are 32 time slots in each frame, numbered 0 to 31. The 0th and 16th time slots are used for framing and signalling information. Time slot 0 contains frame alignment patterns which alternate between successive frames. Some of the bits are not part of the specified pattern and these can be used for data. One bit in each alternate frame is allocated for remote alarm signalling purposes. The first four bits of frame 0, time slot 16, contain the frame align-

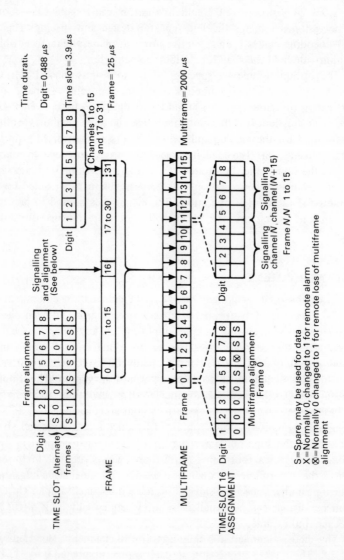

Fig. 7.8 – 30-channel PCM digit assignment.

ment pattern of four consecutive zeros. Three of the other four bits are spare and can be used for data. The eighth bit is used to signal remote loss of multi-frame alignment. Time slot 16 in the remaining 15 frames is used for channel signalling as shown in Fig. 7.8. The four bits per channel give the possibility of 15 signalling states. The all-zero combination is never used to avoid confusion with the multi-frame alignment pattern. This signalling arrangement, where specific bits are allocated for specific channel signalling, is known as channel-associated signalling.

With modern digital switching systems it is possible to make more efficient use of the signalling channel by using it simply as a digital transmission facility, the format of the signalling information being the responsibility of the exchange equipment. This technique is known as common channel signalling. The CCITT common channel signalling standard for use with digital switching networks is known as signalling system No. 7. The No. 7 signalling system operates on a data packet concept with bits allocated for flag, acknowledgement, addressing and error checking. The content of the information field varies according to the particular digital system in use. Unfortunately, to make a more detailed study of common channel signalling would require a much more in-depth consideration of digital switching systems than it has been possible to include in this book.

DIGITAL MULTIPLEXING HIERARCHY

As with FDM, a hierarchy of digital multiplexing has been agreed that enables multiple groups of channels to be conveyed over long-distance, high-quality digital links. Unfortunately differing standards exist in various parts of the world. The CCITT standard in use in Europe is given in Table 7.2. At each stage the number of channels multiplexed is multiplied by 4. Each stage of multiplexing also requires further digits to indicate framing and to carry network control and signalling information. Thus the specified digit rates increase by a factor greater than 4. The rates are chosen to be multiples of 64 Kbits/s, so that the standard octet structure can be maintained even for the higher levels of the hierarchical structure.

Table 7.2 — PCM multiplexing hierarchy.

Hierarchy level	Transmission rate (Mbits/s)	Number of telephone channels
1	2.048	30
2	8.448	120
3	34.368	480
4	139.264[a]	1920
5	565	7680

[a]Nominally 140 Mbits/s.

DIGITAL SWITCHING SYSTEMS

A digitally encoded telephone conversation entering a telephone exchange will occupy a particular time-slot on a particular transmission circuit. On leaving the exchange, it will generally occupy not only a different circuit but also a different time-slot on the outgoing circuit to that which it occupied on the incoming circuit. It will thus require both circuit-switching and time-slot re-allocation. Both of these functions are effectively switching operations. The circuit-switching operation is often called 'space-switching' as opposed to 'time switching' for the time-slot re-allocation function. The calls occupying the various time-slots on an incoming trunk may leave on different outgoing trunks. The space switching may thus need to be changed for consecutive time samples. The space switching of digital signals therefore requires fast switching techniques and electronically implemented logic circuits form a suitable switching medium. Time switching really involves the writing of the time sample into storage and reading it out again at the appropriate time.

A digital switching system can be considered in terms of the equivalent switching networks discussed in Chapter 2. Let us consider the equivalent of a three-stage switching network as shown in Fig. 2.10. This could be implemented in one of two possible ways. These are a sequence of time-space-time switching (TST), or space-time-space switching (STS).

TIME-SPACE-TIME SWITCHING

Let us consider x incoming trunks, each with m time-slots. The x primary switches are time switches and each consists of m storage locations which will allow each sample to be delayed enough to be read out in the appropriate output time-slot. The secondary switch consists of an $x \times y$ cross-point matrix in which the cross-points can be operated at a rate of k operations per sampling cycle. It thus operates with k available time-slots. The x input trunks and y output trunks are connected, via the time switches, to the switch matrix input and output terminals respectively. The tertiary switches comprise y switches, each consisting of n storage locations. To set up a call through the switching network a time-slot on the secondary switch is identified which is not already in use to connect either to the appropriate primary or tertiary switch. The sample is then delayed by the primary switch to occur in the selected time-slot. At this instant the appropriate secondary switch contact is activated to connect the sample to the appropriate tertiary stage. The sample is then delayed again by the tertiary switch so that it occupies the correct time-slot in the output trunk. Assuming that the primary and tertiary switches are identical, that is $m = n$ and $x = y$, then non-blocking can be obtained by providing $k = 2n - 1$ time-slots on the secondary switch matrix. However, this means that the secondary switch does not necessarily operate in synchronism with the primary and tertiary switching stages.

SPACE-TIME-SPACE SWITCHING

In this arrangement the primary switch matrix has m input terminals to accommodate m incoming trunks. The appropriate matrix cross-points are selected sequentially for each of the x incoming time-slots. The primary switch matrix outputs are connected to k time switches, each time switch having x storage locations. In this case it is necessary for y to be equal to x, since it is pointless writing samples into store that cannot be read out again and, conversely, samples cannot be read out that have not previously been written into store. The tertiary switch matrix then connects the time switches to the appropriate output trunk for each time-slot interval. Again, for non-blocking $k = 2n - 1$ time switches are required, given $m = n$, as before. Thus TST and STS digital switching systems can be regarded as the direct equivalents of three-stage cross-point switching networks.

The control of the digital switching network is normally carried out by means of a central computer located within the exchange. Thus the network is said to be operated under stored program control (SPC). The control is therefore flexible and easily modified by simply amending the stored program. Thus subscriber numbers can be re-allocated to different circuit connections to effect 'follow me' transfer or call redirection. It is also not necessary that all samples relating to a single call pass along the same route. All that is required is that the samples pass without interruption and in the correct sequence from the caller to the recipient, that is, a virtual circuit exists. It is therefore possible to reconfigure the network whilst calls are in progress. This could lead to considerable economies in network provision since, by re-routing, under-used plant can often be brought into use to relieve congestion on a heavily loaded route. The advent of digital switching, therefore, opens up the possibility of enhanced network facilities and services. It is also possible to use the same transmission and switching equipment for data communication purposes and it is likely that by the turn of the century the whole telecommunications network will be in the form of an integrated services digital network (ISDN). We shall return again to the discussion of ISDN after we have a taken a long look at the growth of data networks in the next chapter.

8

Data networks

The earliest data networks were provided to enable several terminals to access the same central computer. There were two main uses for such a facility. Firstly, in the early 1960s computers were physically large and comparatively expensive. It was therefore desirable for a number of users to be able to make use of a single mainframe computer by access from remote terminals. Secondly, a number of applications were beginning to emerge where a number of geographically separate users required access to a shared database. Among the earliest of these were the airline seat reservation service networks, the first of which began operation in the United States in 1961. Other early users of computer networks were the banks, who used them to enable branches to obtain direct access to centrally stored account information and to remotely update this information as transactions took place.

POLLED NETWORKS

To allow terminals to communicate with a central computer in turn, it was necessary to establish some form of access protocol. These early networks were operated on a 'polled' strategy, where the central computer invites the various data terminals to communicate in turn according to some predetermined rule. There are basically two types of polling, 'roll-call' polling and 'hub' polling. In roll-call polling the terminals are normally in the 'listening' mode waiting to receive signals broadcast from the central computer. Each broadcast message includes a terminal address field. On recognising its address, a terminal becomes active, either to receive a message or to respond to an invitation to transmit. If the terminal has no data to transmit when it receives an invitation, it simply responds with an indication to this effect. The order in which terminals are polled is thus entirely in the control of the central computer. It is not a pre-requisite that terminals be polled sequentially. Some terminals may be given priority by more frequent entries in the roll-call table than others. The roll-call list can be dynamically re-arranged should priority requirements change. The dis-advantage of roll-call polling is that, for very wide area networks, there could be significant delay in consecutive interrogations owing to propagation delays in the network. This can be reduced by the use of hub polling.

Hub polling differs from roll-call polling in that the interrogation, instead of being initiated by the central computer, is initiated by the preceding terminal on the loop. When a terminal has completed its transaction, or if it has no mes-sage to send in response to the invitation, it passes on the polling invitation directly to the next terminal in the sequence. On long-distance networks, the time required to pass the polling information can be greatly reduced by this means. The penalty for this is more equipment at the terminal to provide the extra intelligence necessary to initiate the polling. It is also more difficult to arrange for priority users and to modify the polling sequence than it is for roll-call polling.

CONTENTION

An alternative strategy to polling is that known as contention. In this arrange-ment, terminals indicate when they wish to send information rather than wait to be interrogated as in the polled situation. In contention systems, a terminal wishing to send examines the highway to see whether it is already in use. If it is, it waits for a short period and then checks again. When it finds the highway free, it transmits its data and awaits acknowledgement. The wait period is necessary rather than a continuous monitoring of the line as, if several terminals are waiting to send, on detecting the end of transmission, all the waiting terminals would commence to send simultaneously, resulting in a data collision. Various techniques, involving random wait times and exponentially increasing waiting periods, have been proposed for avoiding recurrent collisions on heavily used highways. Collisions may be detected by highway monitoring or by non-receipt of message acknowledgements, indicating that the received data has been corrup-ted by simultaneous transmissions.

LOCAL AREA NETWORKS

Much of the requirement for data transmission is for systems confined within the boundaries of a large building or campus. Such systems do not have to rely on services provided by the telecommunications administration (PTT). They are

therefore normally privately owned and designed to meet the special needs of
the user concerned. Such a facility is generally designated a Local Area Network
(LAN) as opposed to a Wide Area Network based mainly on circuits and services
provided by the PTT. A variety of network architectures have been proposed
from time to time to meet the different needs of various users. These can be
classified according to the network topology.

STAR NETWORKS

A star network is based on a central hub which effects the network control
(see Fig. 8.1(a)). The terminals are normally connected by separate circuits to
the hub, although multiplexing may sometimes be used to provide multiple
circuits on a single physical pair. An example of a star network is one based
upon the PABX. Star networks are normally circuit-switched and provide transparent half or full-duplex transmission. Because of the circuit transparency, the
network is unable to carry out code, speed or protocol conversion. They can,
however, generally be used for carrying voice traffic as well as digital communication.

LOOPS

In a loop system the devices are connected to a common highway which takes
the form of a closed loop. One station normally acts as master and polls the
other stations in turn, requesting them to transmit if they have data to send
(see Fig. 8.1(b)). The loop network has two basic disadvantages. Firstly, the
polling strategy means that the response time is relatively slow compared to
other configurations. Secondly, the basic master/slave relationship places unnecessary restrictions on network operation.

RINGS

In the ring architecture, the devices are again connected in a 'loop' but all the
devices now participate in the control strategy. The interconnection of the devices may be either common highway or by interconnection of adjacent devices,
the devices themselves forming part of the transmission ring (see Fig. 8.1(c)).
Several strategies exist for inserting new data into the highway. One method is
register insertion, where data on the highway is temporarily buffered and new
data is inserted ahead of it. Another technique is the 'token passing' method,
where a special code follows the most recent transmission on the ring and a terminal wanting to transmit waits until it detects the token before adding its
message to the highway. It then puts the token at the end of its message. A further method is 'empty slot', in which the highway is effectively time assigned
into separate slots and a terminal waits until it detects an empty slot before
transmitting. Probably the best-known ring topology in the UK is the 'Cambridge Ring'. This employs the empty slot control strategy. Rings are, however,

vulnerable to catastrophic failure and it is thus normally necessary to incorporate a by-pass feature to shunt errant repeaters.

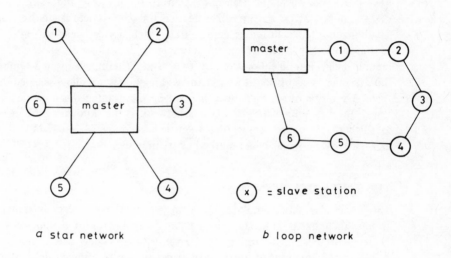

a star network   *b* loop network

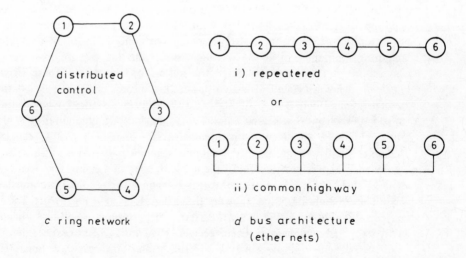

c ring network   *d* bus architecture
                                          (ether nets)

Fig. 8.1 – Local area network configurations.

BUS CONFIGURATION

The bus architecture implies an open-ended common highway topology, as opposed to the closed arrangement of the loop network (see Fig. 8.1(d)). To avoid simultaneous use of the highway by two or more senders, a technique known as CSMA/CD (carrier sense, multiple access with collision detection) is used. Terminals wishing to transmit 'listen' to the highway and transmit only if there is no signal on the line. Should two or more devices decide to transmit simultaneously, a collision is detected and each device backs off, attempting to transmit again after randomly determined time intervals. Because of the random delay, the likelihood of any two or more terminals again attempting a simultaneous transmission is small. This strategy does not allow selected devices to have priority over others and this can be disadvantageous in some applications. The best-known CSMA/CD bus network is that known as 'Ethernet', in fact Ethernet is so well-known that the title is becoming widely used to describe all CSMA/CD networks based on the bus architecture.

THE CAMBRIDGE RING

The Cambridge Ring employs the empty slot strategy. Data is transmitted around the ring at base-band, that is, the signals are not modulated onto a carrier as is the case for broad-band systems. A line code is used, however, to ensure transitions occur in the data sequence so that bit-timing can be recovered by the recipient. The line code used with the Cambridge Ring is the Wal 1 or Manchester Code, described in Chapter 6. The bit transmission rate is 10 Mbits/s. The slot frame consists of 38 bits. There must be sufficient transmission delay around the ring for it to contain at least one whole frame and a couple of extra bits to separate the first and last bits around the ring. There can, of course, be a multiple of frames, each separated by at least two bits from each other. In practice Cambridge Rings almost always operate with only a single slot circulating, requiring a minimum of 40 bit locations around the ring. Bits are observed by a station one at a time, being held at the station for 1-bit period. There is thus a single-bit delay associated with each station. If the connections between stations are fairly long, there is a possibility of further bits being stored owing to transmission delays, but this is not common. Any additional delay required to achieve the minimum of 40 bits stored is provided by the insertion of a shift register in the ring, usually at the monitor station. The monitor station is not a network terminal. It is provided to monitor the network operation and to restore the slot pattern should this be disturbed by transmission errors or faulty terminals. A layout of a Cambridge Ring, showing the delays, is given in Fig. 8.2(a). The slot bit pattern is shown in Fig. 8.2(b). Pattern synchronism is established when the network is first switched on. Each receiver thus knows when to expect the start bit. The start bit is always binary 1 and the surplus bits are always binary 0. It is thus a simple matter to check each time whether synchronism has been affected by bit deletion or insertion, providing this is contained within reasonable limits.

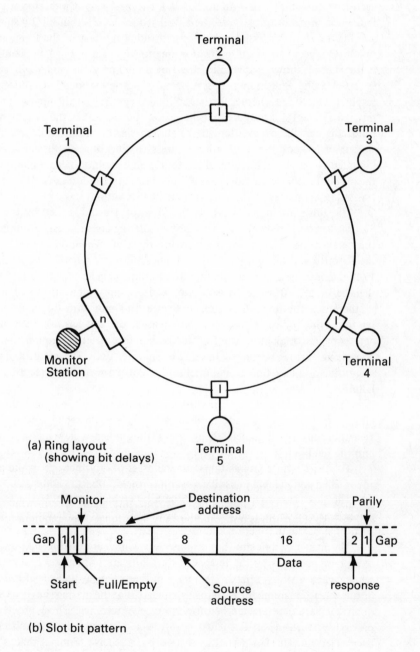

(a) Ring layout
(showing bit delays)

(b) Slot bit pattern

Fig. 8.2 – Details of Cambridge ring.

The second bit in the frame indicates whether the slot is empty or full according to whether it contains a 0 or a 1 respectively. On receipt of an empty indication, a station may seize the slot if it has data to transmit by substituting a 1 for the 0 in the Full/Empty slot and filling the rest of the frame as appropriate. On receipt of a full indication one of three actions can be taken. Firstly, if the station knows that it filled the slot on the previous pass, it will substitute 0 for 1 in the Full/Empty and adjacent Monitor bit locations and check the response bits to see whether further action is necessary. It then passes on the slot marked as empty, signifying that the rest of the bits left in the slot are no longer valid and may be overwritten. If the station was not responsible for filling the slot, it will check the destination address to find out whether the message is intended for it. If so, it will read the source address and the data bits and update the response bits as appropriate. If the destination address is that of another station, it just passes the bits on without further action.

The third bit in the frame is the Monitor bit and is used by the Monitor Station to check that messages are not circulating continuously around the ring, thus preventing other users from seizing the slot. Whenever a 'full' slot passes the monitor station, the monitor bit is checked to see whether it contains a 0 or 1. If it contains a 0, then the monitor station replaces it with a 1. This 1 will then be reset to 0 when the slot is marked as empty on return to its originator. If, therefore, the monitor station receives a 'full' slot with the monitor bit also set to 1, this indicates that, for some reason, the slot has not been marked as empty by its originator. The full slot is therefore rotating around the ring and preventing further seizure of the slot by another user. The monitor station has therefore to take action to re-initialise the slot pattern and to record a network failure.

The next group of eight bits (byte) contains the destination address of any message in the data field and is filled in by the station seizing the slot for message transmission. It is deleted by overwriting when a new message is written into the field.

The next byte contains the source address of any message in the data field and is filled and deleted in similar fashion to the destination address.

The next two bytes contain 16 data bits which constitute the text of the message to be transmitted. They are, again, filled and deleted in a manner similar to that of the destination address.

The next two bits are the response bits, which are used to acknowledge the receipt of messages by the recipient. Both bits are normally set to zero by the sender when a slot is seized. The recipient then changes the pair of bits into one of the three remaining bit patterns to indicate that the message has been accepted, or that it has been rejected because the receive terminal is busy processing data previously received, or that it has been rejected because the transmitter has not been selected. The last response is used by a receive terminal that is awaiting further packets of data which form a single complete message. This is used when a single data slot is insufficient to hold a complete message sequence. When the transmitter receives the response bits, it will then know whether the message has been received or whether it has to retransmit a rejected message. If the response bits have not been altered by the recipient, the transmitter will deduce that there is no active mode on the network with that address.

The final bit in the frame is a parity bit. The parity of the message is checked at each node and failure of the parity check is noted. Repeated failure of parity may be taken as an indication of a failure of the preceding node. Various actions can be taken on detection of parity failures depending on the network user requirements. If the monitor station has diagnostic capabilities, it is usual to pass the appropriate information on to it to initiate further action.

ETHERNET

Ethernet is a CSMA/CD bus network. The stations are connected to a common bus highway. The network is therefore passive; that is signals pass directly from the sender to the receiver. They do not pass through the other stations or any other repeaters as they generally do in the ring strategies. Strategies where the network requires signals to be regenerated as they pass around are known as active networks. The Cambridge Ring is therefore an example of an active network. Transmission is at 10 Mbits/s and base-band transmission is used. Since the network is passive, the network is not intrinsically synchronous and it is therefore necessary to synchronise the sending and receiving terminals each time a transmission path is established. Each packet, therefore, has to commence with a synchronising preamble to ensure synchronism is achieved before the transmission of meaningful data commences. The preamble normally consists of a sequence of 32 bits. This is followed by the Destination and Source Address fields. Because of the possibility of interworking of local Ethernets, it is considered to be desirable that terminal addresses should be universally unique, rather than unique only within the confines of the local network. Each of the address fields therefore consists of 48 bits, which gives a possibility of over 280 billion (2.8×10^{14}) different addresses. The next 16 bits are used to define the packet type. Then follows the data field, whch can vary to anything between 45 and 1500 bytes, depending on the individual message requirement. A minimum packet size is necessary to ensure satisfactory operation of the contention procedure. For messages requiring less than 45 bytes, the data field must be filled with dummy data up to the minimum size. The maximum field size is necessary to prevent a single user monopolising the transmission facility. On Ethernet, therefore, a complete message usually only occupies one packet, whereas in the Cambridge Ring it is unusual, except for acknowledgement messages, for a message to be contained within a single packet.

The data field is followed by a Check-sum field of 32 bits produced using a Cyclic Redundancy Check-sum Code over the address and data fields. The basic Ethernet frame field is shown in Fig. 8.3.

When a station wishes to transmit, it monitors the line to see whether transmission by another station is in progress. If so, the station withdraws and waits for a short period before contending for the highway again. The small delay is

Sync	Destination address	Source address	Packet type	Data	Check-sum
32 bits	48 bits	48 bits	16 bits	Variable 45-1500 bytes	32 bits

Fig. 8.3 – Ethernet frame field allocation.

necessary to prevent several waiting stations attempting to acquire the bus simultaneously when transmission ceases. If the bus is clear, the station begins transmitting its synchronising pattern. However, it continues to monitor the line for a further period equivalent to the longest transmission delay likely to be encountered on the bus. In this way the station can detect any collisions that might occur owing to two (or more) stations detecting the same silence and both commencing to transmit before receiving data already sent by the other terminal but delayed in propagation along the bus.

The maximum length for a single Ethernet bus is about 500 metres. For lengths greater than this, repeaters are required. Up to a 100 tap transceivers can be connected on to one Ethernet segment. The transceiver connects the user

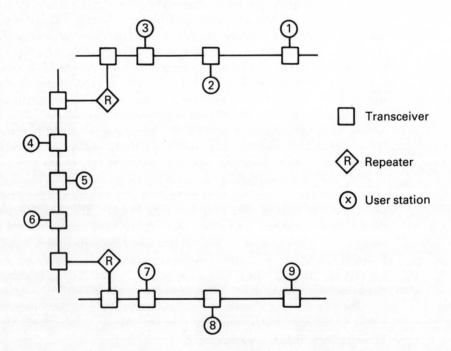

Fig. 8.4 – Multi-segment Ethernet configuration.

station interface cable to the common bus highway. Larger systems consist of several segments interconnected via repeaters. The repeaters act as an interchange between the two interconnected segments. Such an arrangement is illustrated in Fig. 8.4.

WIDE AREA NETWORKS

Wide Area Networks (WANs) are those networks which go beyond the bounds of the building or campus and therefore have to make use of the facilities of the public telecommunications carrier to interconnect the various sites involved. There are available three basic types of facility for this application. Firstly, the public telephone network can be used to carry data by means of the Datel services described in the previous chapter. This provides transparent data links, at data rates adequate enough to to support communication with a visual display unit. However, the data rates are several orders of magnitude slower than those used within Local Area Networks. Secondly, providing the necessary access to the appropriate exchange can be arranged, then PCM telephone channels can be acquired for data-transmission purposes, providing a data rate of 64 Kbits/s. British Telecom offer this facility under the trade description 'Kilostream'. It is also possible, if required, to acquire a complete 30-channel PCM group, giving a data transmission capability of 2.048 Mbits/s. British Telecom offer this under the name 'Megastream'. These have to be rented as leased transmission facilities and there is no capability for circuit switching within the facility. With the ending of the British Telecom monopoly, other carriers are now beginning to offer a similar facility, mainly based on optical fibre links.

The third, and by far the most flexible, facility available is the packet-switched network. Packet-switched networks now exist in most European countries, throughout the United States and in many other parts of the world. They offer a flexible switched data facility for a wide variety of data applications. An almost universally accepted interface standard has been adopted which allows data terminals to be interconnected by the network in an unrestricted way. This concept of unrestricted interconnection is known as 'Open Systems Interconnection' (OSI). To achieve Open Systems Interconnection a hierarchical protocol structure has been proposed by the International Standards Organisation known as the ISO 7-layer protocol model for OSI. Before we look in more detail at packet switching it will be useful to take a quick look at the OSI model.

OSI PROTOCOL MODEL

The OSI model identifies seven levels or layers for the definition of protocols and interfaces for open systems of interconnection. The model is illustrated in Fig. 8.5. At the lowest level, level 1, physical parameters for signals are defined simply to enable data signals to be transferred over the physical connection. This layer, the Physical Layer, consists of specifications for the line code, transmis-

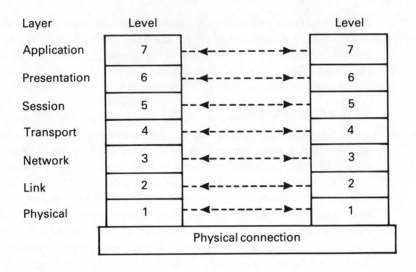

Fig. 8.5 — The ISO 7-level model for open systems interconnection.

sion rate, signal voltage levels, physical connectors and other parameters to enable satisfactory transfer of streams of digits over a simple single connection path. At this level, no significance is attached to any single digit, it is concerned only with the efficient transfer of data across the physical connection.

The second layer, the Link Layer, is concerned with the establishment of a disciplined and reliable data link across the physical link. At the physical level there is an inherent degree of unreliability in that there is no knowing whether errors have occurred in the transfer of data. Thus one of the major functions of the Link Level is to provide error detecting and/or correcting facilities. This in itself may require the division of the data stream into blocks so as to identify the field over which any error-detecting code operates. Alternatively, block delimiters may be necessary for synchronisation purposes. Although various digits now perform specific functions, nevertheless the basic transparency of the data link is maintained in the bits allocated to the data field in the Link protocol.

A data network comprises a number of nodes interconnected by various data links. The third layer in the OSI model is concerned with network operation and is known as the Network Layer. At this level the addresses of terminal nodes are included, together with control and acknowledgement fields to ensure that data communication is correctly established between the appropriate network nodes. The Network control should be independent both of the Link Level of control and the higher levels of protocol.

The three layers so far considered are concerned only with the communications network and are thus properly the concern of this book. The fourth layer, the Transport Layer, takes into account the nature of the terminal equipment and is concerned with establishing a transport service suited to the needs of this equipment. It must thus select a link through the network which operates at a data rate and quality appropriate to the needs of the terminals involved in the communication operation. We are thus beginning to depart from the basic task of providing a communication path through the network.

The next three layers, layers 5 to 7, are task-oriented and have to do with the operations performed by the data terminal equipment rather than with the network. The Session Layer is concerned with setting up an operational session between terminals. It can thus be basically identified with the operation of 'signing on' the computer to begin the operation of the desired task. The Presentation Layer is concerned with the format in which data is to be presented to the terminals. The Application Layer defines the nature of the task to be performed. These three higher order layers are mainly concerned with the organisation of the terminal software and are not directly the concern of the communications engineer. The Transport Layer is the layer which links the communication processes to these software-orientated protocols.

The basic philosophy of the 7-layer model is that each layer may be defined independently of every other layer. Thus from a user point of view, interchange takes effect across each layer as shown by the broken line connections in Fig. 8.5. In fact each operation passes down through the layers of the model until data interchange is effected through the physical connection.

We shall now look in more detail at packet switching and then we shall return to the OSI model to see how levels 1 to 3 are implemented in a packet-switched network.

PACKET-SWITCHED NETWORKS

In circuit-switched networks a circuit is allocated for the whole duration of a call. The proportion of time during which data is actually being transmitted over the circuit is often quite small and thus far from efficient use is made of the network facilities. Much more efficient use of the network can be obtained if the network circuits are used to interconnect network nodes and that data is passed from one node to another in a store-and-forward mode of operation. The data messages are divided up into suitably dimensioned data packets, which enter the network through one of the nodes and are passed on from node to node until they reach the node which serves the destination terminal. Packets forming part of the same message do not necessarily take the same route through the network or utilise the same circuits. However, as far as the sending and receiving terminals are concerned, a 'virtual circuit' is established between them so that they are generally unaware that the network is anything but circuit-switched. The network nodes are thus connected together by a logical channel rather than a direct circuit path. The choice of path through the network for each packet is determined by the traffic on the network at the time the packet enters the network. Because the traffic will be constantly changing, packets which form part of the same message may be routed through different nodes and circuits and may experience different delays in the store-and-forward procedure. In normal operation, therefore, it is possible for packets forming part of the same message to arrive at the receiving node in a different sequence to that in which they were entered at the sending node. In applications where the correct sequence is important, all packets have to be delayed at the receive node sufficiently long for the packets to be resequenced before being passed on to the receive terminal. Alternatively, once a virtual call has been established, the same logical circuit can be used for all packets associated with the same message. Where the packet sequence is unimportant, the packets can be regarded as independent entities, usually referred to as 'datagrams'. In datagram operation each packet has to contain both source and destination addresses since, being an independent entity, it cannot make use of the information contained in the packets associated with setting-up the virtual call. Thus, although datagram operation makes more efficient use of the network transmission facilities, it requires a larger packet overhead than would be required if all the packets associated with a single message follow the same route once the virtual circuit has been established.

HDLC AND X25

The OSI model level 3 protocol generally used with packet-switched networks is that defined in CCITT recommendation X25. X25 operates within a level 2 protocol known as High-level Data Link Control (HDLC). This in turn utilises the physical interface specified in CCITT recommendations X21 or X21 bis. X21 bis is compatible with CCITT recommendation V24 for interfaces for modems operating over the public telephone network. It should perhaps be pointed out here that V24 defines the data terminal equipment interface to the modem and is not concerned with the interface between the modem and the line itself. Thus the modems themselves form part of the conceptual physical connection. The V24 interface is thus independent of both modulation technique and data throughput rate.

The HDLC format is shown in Fig. 8.6. The X21 data stream is divided into frames being delimited by a flag sequence consisting of eight bits. The flag octet used in HDLC consists of the bit pattern 01111110. To prevent false flag indications occurring due to the occurrence of the flag octet pattern in the rest of the

Flag	Add.	Cont.	Information field (X25)	Frame check sequence	Flag
8 bits	8 bits	8 bits	Variable	16 bits	8 bits

Fig. 8.6 – HDLC frame structure.

frame field, a technique known as 'bit stuffing' is used. Whenever five consecutive 1s occur in the data, an extra 0 is inserted (stuffed) into the digit stream following the fifth 1 digit, whatever the next digit is. Thus the sequence of six 1 digits is unique to the flag. Because a 0 is stuffed whether the next digit is a 0 or 1, on receipt of five consecutive 1s it is a simple matter to remove the next 0 from the digit sequence, thus reverting to the original binary data packet content. Because the size of the HDLC frame is in any case variable, the packet format is not impaired by the insertion of occasional extra 0 digits. The bit stuffing is performed on the complete packet format and the basic packet structure is performed before the bit stuffing operation is carried out.

The HDLC packet proper comprises four fields. The first field consists of 8 bits known as the Address field. The address specified in this field is that of the secondary node, that is, the node corresponding to the called party. The node associated with the calling party is known as the primary node. Thus all communication processes originate from the primary. Thus messages originating from the primary contain the address of the node to which the message is directed. Response messages, however, contain the address of the node from which the response originates. Because of the network strategy, the other address is implicit in the interchange procedure and the single address field is sufficient for network housekeeping purposes.

The second field consists of a further 8 bits and is referred to as the Control field. The significance of the bits depend on the packet type and certain of the bits are actually used to define the type of packet. The significance of the bits for each packet type is given in Fig. 8.7. If the first bit is a 0, then the packet is an 'information' frame, that is, it is the normal type of frame used when there is information to be transmitted. In this case, bits 3 to 5 and 6 to 8 are sequence check numbers $N(S)$ and $N(R)$. Packet switching does not rely on semi-permanent circuits being established for the whole duration of the call. Instead, packets are routed individually through the network by any suitably available link. Since consecutive packets may take different routes, with different delay factors, it is possible for packets to arrive at the receiving node in a different

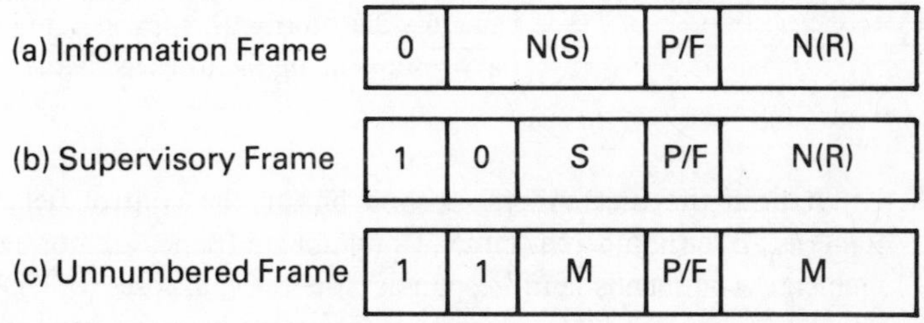

(a) Information Frame	0		N(S)	P/F	N(R)	
(b) Supervisory Frame	1	0	S	P/F	N(R)	
(c) Unnumbered Frame	1	1	M	P/F	M	

Fig. 8.7 – HDLC control field structure.

sequence to that in which they were transmitted. To ensure the receiver is aware of this, each succeeding packet is numbered consecutively in binary notation modulo-8 in the three digit slots designated N(S). Since differential delays in excess of eight-packet periods are highly unlikely, the use of modulo-8 counting is generally adequate for all practical purposes. The sequence numbering in the N(S) sub-field is carried out independently by both the primary and the secondary. The N(R) field is used to indicate the number of the next packet in sequence that the node expects to receive and therefore acts as an acknowledgement to the sender that the previous packet has been correctly received. The sender can thus keep tabs on which of his message packets have been received and decide whether action is necessary to retransmit any messages received in error or not received at all.

If the first bit of the Control field is 1, then the second bit also defines the packet type. If the second bit is 0, then the frame is a supervisory frame. The supervisory frame is used to send the N(R) acknowledgement when there is no information to be sent on which the acknowledgement can be sent piggy-back'. As there is no information to be sent, there is no sequence number N(S) to be sent. The other two bits, shown as S in Fig. 8.7(b), can then be used to signal the state of the sending terminal. The two bits can be used to signal four states. These are:

(a) Receive Ready: Indicates that the node has correctly received a packet and that it is ready to recieve the next packet in sequence which it expects to be number N(S).

(b) Receive Not Ready: Indicates that the node has correctly received a packet and that the next packet it expects to receive is number N(S). However, it also indicates that the node is temporarily not ready to receive it.

(c) Reject: Indicates that a packet has been received in error. N(S) will therefore indicate the number of the packet from which retransmission should commence.

(d) Selective Reject;   Indicates a specific packet has been received in error and requests retransmission of that packet only. This can only be used with sophisticated networks which have resequencing facilities available.

If both the first and the second bits in the control field are 1, then the frame is an 'unnumbered frame'. Unnumbered frames are normally used to communicate commands and responses for the purposes of link housekeeping rather than for the interchange of information between network nodes. As they are not normally associated with the transmission of information packets, there is no requirement for packet sequence numbers. Bits 3,4,6,7 and 8, designated M in Fig. 8.7(c), are thus all available for command and response signals. Thus 32 different signalling combinations can be specified using the M field. The list is complex and only of specialised interest. The reader is directed to the CCITT recommendation X25 or one of the specialised books on data networks if further details are required.

Whatever the type of Control field, the fifth bit is always what is known as the Poll/Final bit, designated P/F in Figs. 8.7 (a), (b) and (c). In frames originated by the primary the P/F bit is set to 1 to poll the secondary, that is, to instruct the secondary to issue a response. In frames originated by the secondary the P/F bit is set to 1 to indicate to the primary that it is sending the final frame in a message sequence.

The HDLC Information field carries the X25 data packet which is itself structured to control the level 3 network functions. We will defer further discussion of the X25 format until we have completed our discussion of HDLC.

The final field in the HDLC frame is the frame check sequence. This consists of 16 bits obtained using a cyclic redundancy check code over the Address, Control and Information fields of the HDLC frame. The characteristic polynomial used in HDLC is the CCITT V41 polynomial

$$X^{16} + X^{12} + X^5 + 1$$

as described earlier in Chapter 5. The frame check sequence is generated before bit stuffing is carried out.

The X25 level 3 packet occupies the Information field in the HDLC level 2 frame. The higher order of control exercised at level 3 means that the X25 packet structure is more complex than that of HDLC and there is a much wider range of packet types. Space does not permit a full discussion of each type of X25 packet. We shall therefore have to content ourselves with a look at the general packet structure so that we can at least appreciate the general philosophy of X25 operation. A generalised format structure for X25 packets is given in

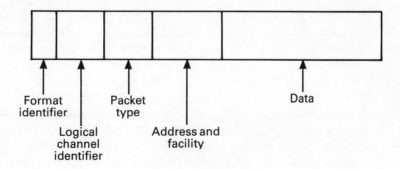

Fig. 8.8 – Generalised X25 format structure.

Fig. 8.8. The packet firstly contains a Format Identifier which defines the format of the rest of the packet fields. This is always followed by a Logical Channel Identifier which carries the number of the logical channel. Each call is allocated a logical channel number so that packets related to the same call can be identified. This is followed by a Packet Type field which specifies the function of the packet.

Packets can be of several types. Packets involved in call initiation will contain full address information in the next field slot. Subsequent packets do not require full address information since the packet routing is now under the control of the Logical Channel Identifier. The address/facility field can then be utilised for sequence numbering or acknowledgement information. Acknowledgement and Control packets do not require a data field. Information packets will contain a data field of variable length, usually with a maximum size of 128 octets.

THE PROVISION OF PACKET-SWITCHED NETWORKS

Packet switched networks, both for public and private use, exist in most European countries, throughout North America and in many other parts of the world. These are mostly accessed through X25 interfaces. In the UK, an X25 packet-switched network is provided by British Telecom and is accessible to subscribers at least in most of the larger towns and cities. In some cases the access has to be via the public telephone network using Datel modems with a V24 interface. Many of the packet-switched networks can be interconnected to provide international data communication at the X25 level.

There is sometimes a requirement to connect data terminal equipment to the packet-switched network which does not contain sufficient intelligence to operate through an X25 interface. Such terminals normally operate in an asynchronous character mode, providing an interface through a serial character stream. Another possible requirement is the interconnection of local area networks through the public packet-switched network. To perform these functions a Packet Assembler/Disassembler (PAD) is required. The PAD converts the character stream into X25 packets for transmission through the packet-switched network and also converts the X25 packets back into asynchronous character format before passing on to the receiving terminal equipment. Three separate X series CCITT recommendations are associated with the PAD, X3, X28 and X29. We shall not go into details of these recommendations here. Suffice it to say that because of the three separate recommendations relating to the PAD, the standard is often referred to as the 'Triple X' standard. Because PADs may be required for a variety of different terminals and applications, there are several options within the Triple X recommendations and a range of PADs are available for specific applications. More complex general-purpose PADs can also be obtained which can be programmed for several different applications.

INTEGRATED SERVICES DIGITAL NETWORK

The conversion to digital transmission and switching of speech signals in the public telephone network is proceeding at such a rate that it is likely to be almost universal within the next decade or so. The only part of the network which has not to date seen a move towards digital transmission of speech is the local connections between the subscriber and the local exchange. However, with the rapid progress of technology towards a greater use of large-scale integration of circuits onto a single chip, it is now becoming a practical proposition to provide analogue-to-digital and digital-to-analogue conversion in each telephone handset. Work is therefore progressing towards providing digital transmission over the subscriber's line. The result of this will be that high-speed digital transmission facilities will be available from subscriber's apparatus to subscriber's apparatus. It then becomes unnecessary to distinguish between speech and data signals, the same network being available for both without discrimination. This leads us to the concept of an Integrated Services Digital Network (ISDN). There is a commitment to provide such a network in the UK by around the turn of the century. The network will effectively be circuit-switched, though it is likely the X25 format will be used for data terminal access.

The problem with the provision of digital transmission over the subscribers' lines is that the lines vary considerably in quality and length and that they exist as a single pair. Thus the necessary full duplex operation must be obtained using multiplex techniques. Two methods have been considered. The first is known as 'burst mode' or 'ping-pong' operation. In this technique short bursts of digits are sent alternately from each end of the communication path. Because some time is required for propagation of digits along the line, the basic transmission rate has to be significantly greater than the total two-way data throughput rate. Thus basic transmission rates of the order of 250 kbits/s may be necessary. An alternative is to use electronic hybrid circuits where transmission in each direction takes place simultaneoulsy but the signal being transmitted by a given terminal is subtracted from that being received from the line by the same terminal, so leaving only the signal from the other terminal at the receiver input. This means that the basic transmission rate is equal to the single direction data throughput rate, rather less than half that required for burst mode.

The basic rate for PCM speech signals is 64 kbits/s. However, it is proposed to use the local line for simultaneous transmission of digital speech and data. One proposal, therefore, is for 64 kbits/s for speech plus 8 kbits/s for data plus 8 kbits/s for network management and signalling, making a total requirement for 80 kbits/s full duplex over the local line. For most local lines this is just about possible using the burst mode technique. A second, more ambitious, proposal is 64 kbits/s for speech, 64 kbits/s for data, 8 kbits/s for network management and signalling a further 8 kbits/s for a second data facility or for further signalling. This gives a total of 144 kbits/s full duplex over the local line. This will almost certainly require the use of the electronic hybrid technique. Although this technique requires more complex circuitry than burst mode, modern technology now makes the provision of the necessary electronics an economically viable proposition. It is likely that in the next few years ISDN will become the internationally accepted standard for all telecommunications services, offering flexible use of the available transmission facilities for a wide range of applications.

MULTIPLEXORS AND CONCENTRATORS

Many wide area network communication nodes are connected together by long-distance data links. Many of these links are capable of high rates of data transmission in comparison to that required by the data terminal equipment. For example, the Kilostream rate of 64 kbits/s is many times that required by a typical VDU operating at, say, 2.4 kbits/s. To make efficient use of the long-distance transmission facilities it is often possible to combine together signals from a number of different terminals to form a single stream of data. This can then be sent over a single transmission channel, thus making much more efficient use of the available transmission capacity.

There are two ways in which this combined use of a single channel can be implemented. Firstly, it can be achieved by straightforward multiplexing, usually on a time-division basis. The multiplexing may be carried out by taking a bit from each terminal in turn, or a byte or character from each terminal in turn, or indeed any appropriate size of packet to suit the application. The total bit-rate from all the user terminals cannot exceed the channel bit-rate, although terminals can be connected up to this rate. Thus up to 26 2.4 kbits/s terminals can be multiplexed onto a 64 kbits/s channel, leaving some bits available for framing purposes. It is not necessary that all the users operate at the same bit-rate, provided the total capacity is not exceeded. Thus a 64 kbits/s channel could service, say, three 9.6 kbits/s, four 4.8 kbits/s and six 2.4 kbits/s terminals, as shown in Fig. 8.9. It is obviously simpler to arrange the multiplex frame structure if there is a simple arithmetical relationship between the contributing terminal bit rates. At the far end of the communication channel the contributing data streams are separated back into separate circuits for connection to the appropriate data terminal equipment.

In many applications the contributing channels are not always fully occupied. Under these circumstances it is possible, by suitably addressing the multiplex

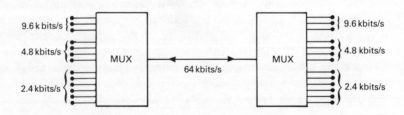

Fig. 8.9 – Multiplexor with assorted input rates.

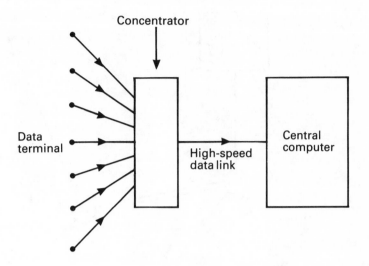

Fig. 8.10 – Concentrator arrangement.

packets, to connect more terminals than simple multiplexing would normally allow. Use can thus be made of the statistical nature of the data sources to make even more efficient use of the communication channel. The multiplexing equipment, known as a statistical multiplexor (STAMUX) is more complex than the conventional multiplexor (MUX) and hence more expensive. However, on long-distance data links it is often possible to recoup the extra equipment cost in the saving in cost achieved by more efficient use of the transmission facility.

Often the data network consists of a number of remote terminals connected to a central computer. It may thus be unnecessary to separate the multiplexed data into separate channels at the point of receipt. Instead, the data can be concentrated at a local point before onward transmission to the central data processor. Thus instead of providing a multiplexor, a concentrator can be used. The concentrator may simply assemble the data in packets in a way analogous to that used in the multiplexor, or it may be itself a miniprocessor that can carry out some preliminary data processing before the data is passed on to the remote mainframe computer. A typical concentrator arrangement is given in Fig. 8.10.

A complex network arrangement may contain a mixture of multiplexors and concentrators, together with a variety of transmission channels and modems. A typical complex wide area network configuration is given in Fig. 8.11. There is an almost infinite variety of ways in which such a network might be configured, each being dependent on the user requirements and available facilities.

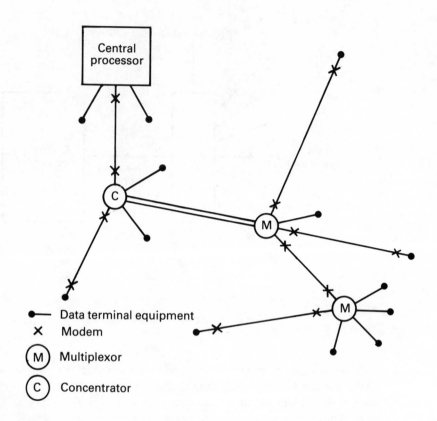

Fig. 8.11 – Typical complex Wide Area Network.

9

Optical fibre transmission

Light is really an extremely high frequency electromagnetic wave. It is therefore possible to modulate it and transmit it like any other electromagnetic wave. Because of its extremely high frequency, between about 2×10^{14} and 5×10^{14} Hz, enormous band-widths are possible and an almost unlimited number of telephone conversations could be modulated into the frequency bandwidth of a single optical fibre. Light is propagated along optical fibres in a way that is analogous to the way in which microwaves are propagated along waveguides. The light is constrained within the transparent fibre by cladding with a material having a lower refractive index than the core material, thus causing total internal reflection. If the core fibre is of sufficiently small diameter, a single mode of transmission is possible.

There are three types of fibre in use, multimode fibre, monomode fibre and graded index fibre. We shall look first at multimode fibre.

MULTIMODE FIBRE

In multimode fibre, the diameter of the fibre core is significantly greater than the wavelength of the light being propagated along the fibre. The light wavelengths are between about 0.6 and 1.6 μm. The core diameter used in multimode fibres is typically 50 μm. This is clad with a layer of low refractive index material to give an overall fibre diameter of about 60 μm. Because the diameter is large compared with the wavelength, it is possible to consider the light propaga-

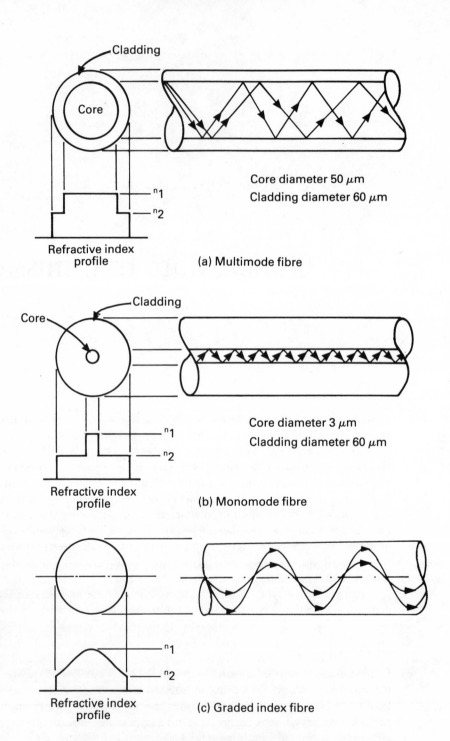

Fig. 9.1 – Optical fibre types.

tiion simply in terms of reflected light rays. The rays propagate along the fibre by means of multiple internal reflections as shown in Fig. 9.1(a). It can be seen that a single pulse of light will travel along a band of paths of differing lengths, depending on the angle of incidence of the light ray at the surface of the fibre. These differing path lengths mean that a light pulse becomes dispersed in time as it propagates along the fibre. This leads to pulse stretching which limits the rate at which consecutive pulses can be transmitted without adjacent pulses interfering with one another. A typical transmission rate over multimode fibres of length 1 km is about 100 Mbits/s. Since information is conveyed by the intensity of the light emission, wide-band, incoherent, light-emitting sources can be used. A simple photo-emitting diode is all that is required. Detection and conversion back into an electrical signal can be carried out using a simple photo-sensitive detector.

MONOMODE FIBRE

In a monomode fibre the diameter of the fibre core is of a size commensurate with that of the wavelength of the light emission. A typical core diameter is 3 μm, usually clad to give an overall diameter of 60 μm, this being about the thinnest fibre that can be handled mechanically without damage. At these dimensions it is no longer valid to consider the propagation in terms of simple reflections of light rays along the fibre. Instead, the propagation has to be considered in terms of a wave front propagating along a tubular waveguide formed by the inner core and the interface between the core and the outer. If the wavelength of the emission and the core diameter is suitably chosen, it is possible to arrange for a single mode of propagation only along the fibre. This effectively gives a single transmission path which in turn leads to low dispersion and hence a high bit-rate capability. Rates of 10 Gbits/s over 1 km fibre sections are typical. The very small physical size of the core makes jointing of cable sections and interface with input and output transducers difficult. However, increasing the physical dimensions to ease this problem will allow multimode transmission to take place. This will lead to greater pulse dispersion and hence low transmission rate capability. The monomode fibre is illustrated in Fig. 9.1(b). To make optimum use of the monomode fibre capability it is necessary to use a single-frequency coherent carrier. This requires the use of a laser to generate such a signal. Detection may still be carried out by the use of photo-sensitive devices, but these have to be more precisely manufactured for high-frequency operation. A monomode cable system is therefore much more expensive to manufacture and install than a multimode fibre. It does, however, have a capability of at least 100 times greater transmission rates.

GRADED INDEX FIBRES

An alternative to the multimode and monomode fibres is the graded index fibre. In this fibre there is no distinct transition from a high refractive index to a low refractive index material. Instead, the refractive index of the fibre material is made to decrease radially from the fibre axis as shown in Fig. 9.1(c). The gradually changing refractive index leads to a continual bending of the light rays until the critical angle is reached and the light is diverted back towards the centre of the fibre. The propagation path is thus as shown in Fig. 9.1(c). The differing path lengths for rays with differing launch angles leads to some dispersion. The transmission performance for graded index fibres therefore lies somewhere between that of multimode and that of monomode. Likewise, installation and jointing problems are easier than for monomode but worse than for multimode fibres. They thus offer the possibility of a compromise solution, providing production difficulties in obtaining a suitable index gradient can be overcome.

LOSSES IN OPTICAL FIBRES

There are a number of loss mechanisms in optical fibres which cause attenuation of light signals at various frequencies in the spectrum. In general, fibres have a spectral 'window' between wavelengths of about 0.6 μm and 1.6 μm, where the attenuation is at a minimum. A typical curve of attenuation against wavelength is given in Fig. 9.2. At the longer wavelengths, in the infra-red region of the spectrum, there is significant attenuation due to the material absorption of

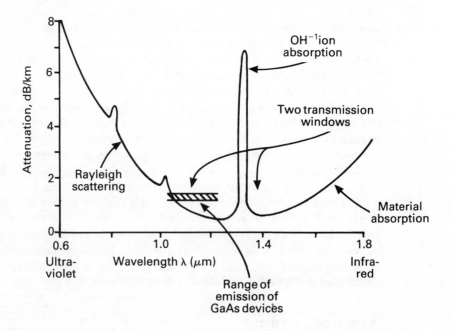

Fig. 9.2 – Typical attenuation characteristic for optical fibre.

energy caused by molecular vibrations in the crystal structure. At the shorter wavelengths, in the ultra-violet region of the spectrum, there is again some attenuation due to absorption of energy, this time caused by ion excitation. However, a much more significant cause of attenuation in this region is that caused by Rayleigh scattering. Rayleigh scattering is caused by irregularities in the crystal structure of the fibre material and is roughly proportional to λ^{-4}, where λ is the wavelength of the light transmitted. Within the 'window' between these two regions of attenuation there are further absorption peaks due to the presence of hydroxyl (OH^{-1}) ions caused by water contamination of the fibre material in the forming process. This causes a large attentuation peak in the middle of the low attenuation region, with smaller peaks at associated resonances in the ultra-violet region. The most common source for high-grade fibre systems is the gallium-arsenide laser, and this, fortunately, together with the gallium-arsenide light-emitting diode, radiates in a region just below the main OH^{-1} absorption peak.

AN OPTICAL FIBRE COMMUNICATION SYSTEM

The key components of an optical fibre communication system are shown in Fig. 9.3. We shall look briefly at each component in turn.

(a) Transmission medium

The choice of transmission medium is between multimode, monomode or graded index fibre. The choice is mainly dictated by the data-rate requirements

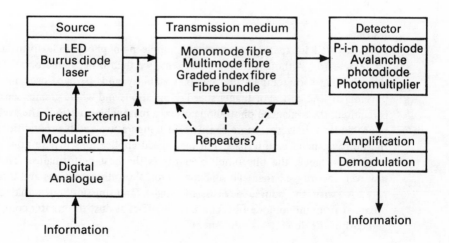

Fig. 9.3 – Block diagram of optical fibre communication system.

over a single fibre. For data-rates up to about 10 Mbits/s a multimode fibre is normally used because of the ease of terminating and the lower cost per metre of the fibre itself. In many applications it may be preferable, and cheaper, to obtain higher transmission rates by using a bundle of multimode fibres rather than a single monomode fibre. Certainly for such applications as digital telephony, security of service may require the greater reliability associated with several multimode fibres rather than committing several thousand telephone calls to one single fibre strand no thicker than a human hair. There are many applications, for example digital television, where bit rates in excesss of 10 Mbits/s are highly desirable. Monomode fibres offer a possible solution. The choice of transmission medium is therefore a trade-off between cost, reliability, ease of installation and maintenance and transmission rate requirements.

(b) Source
The choice of source is mainly dictated by the choice of medium. For low bit-rate systems using multimode fibre, the reasonably wide-band non-coherent light emitted from the light-emitting diode (LED) is normally quite adequate. The LED is a p-n junction device which emits light in accordance with the forward bias current through the junction. The light output is reasonably proportional to the diode current and hence modulation is a fairly straightforward procedure.

For high-speed systems using monomode fibre, a coherent light source is necessary. For this we require a laser, the gallium-arsenide laser being the most widely used. Lasers are considerably more expensive than light-emitting diodes and require more expensive drive circuits. They are also less reliable than the ligh-emitting diode and therefore require more frequent servicing and replacement.

(c) Detector
For low-speed inexpensive systems the simple p-i-n photodiode forms an adequate detector. Although they are less sensitive than other detectors, they have the advantages of temperature and bias stability and, perhaps even more important, of low cost. For higher-speed applications and where greater sensitivity is required, the avalanche photodiode offers a reasonable solution. The avalanche photodiode is, however, more sensitive to temperature and bias conditions than the non-avalanche counterpart. For high-speed applications, where noise factor is of importance, the photomultiplier offers the best performance. They are, however, less rugged than the semiconductor photodiode devices and less compact. As with the source devices, the higher transmission rates that can be obtained from monomode fibres have to be offset against the greater complexity and cost of the detection requirements.

(d) Modulation
The light source is usually modulated by variation of the current through the device junction. The light emitted from the junction is roughly proportional to

the device current, thus it is possible to perform analogue modulation in this way. Digital modulation, using binary signalling, can be performed by switching the current between zero, when there will be no emission, and the value of current giving maximum emission without junction overload. In practice the current is never reduced to zero since faster switching, and hence a higher modulation rate, can be achieved if a minimal flow of current is maintained.

In digital systems it is also possible to modulate the light beam externally to the emitting device. The light beam is deflected opto-electronically so that it no longer couples into the fibre, so producing the zero transmission signal state. Although the technique is possible, suitable devices have not yet been fully developed and direct electrical modulation is the preferred technique. External modulation of the light beam does not appear to offer any feasible technique for analogue signal modulation.

(e) Demodulation
The output current from the light detector is roughly proportional to the power of the light signal impinging on the photosensitive junction of the detector device. This signal is then amplified and processed to yield an estimate in the receiver of the transmitted signal.

(f) Repeaters
Repeaters, which normally involve detection and retransmission of the information-bearing light beam, may be required, depending on the length and loss coefficient of the fibre connection. Typical repeater spacings for various digital modulation rates are given in Table 9.1.

OVERALL LOSS IN AN OPTICAL FIBRE SYSTEM
Generally the output power P_{ot}, launched from an LED into an optical fibre is approximately proportional to the LED input current I_t. Thus

$$P_{ot} = K_L I_t \ , \tag{9.1}$$

where K_L is a constant dependent on the efficiency of the LED and its coupling into the fibre. If the LED is modulated by superimposing a signal current on a d.c. bias, then the electrical signal power input

$$P_t = i_t^2 R_D \ ,$$

where i_t is the rms value of the modulating signal current and R_D is the slope resistance of the LED. Hence, from (9.1),

$$P_t = \left(\frac{P'_{ot}}{K_L}\right)^2 R_D$$

where P'_{ot} is the signal-bearing component of the optical power launched into the fibre.

Table 9.1 – Typical repeater spacings for optical fibre systems.

System	Information rate (Mbits/s)	Repeater spacing (km)
LED and multimode optical fibre of loss 3 dB/km	2.048	10.0
	8.448	6.0
	34.368	1.7
Laser and monomode optical fibre of loss 3 dB/km	2.048	15
	8.448	13
	34.368	10.4
	139.264	8.5
	565	6.5

The reverse current I_r in a photodiode is proportional to the incident optical power P_{or}. Thus

$$I_r = K_d P_{or} ,\qquad (9.2)$$

where K_d is a constant dependent on the sensitivity of the photodiode and the efficiency of the coupling to the fibre.

If the photodiode current is fed into a load resistance R_L and the rms value of the signal component of the photodiode current is i_r, then the electrical signal power in the load

$$P_r = (K_d P'_{or})^2 R_b$$

where P'_{or} is the signal-bearing component of the optical power incident on the photodiode.

The overall system loss

$$y = 10 \log_{10} \frac{P_t}{P_r} \text{ dB}$$

$$= 10 \log_{10} \frac{P_t}{P'_{ot}} \left(\frac{P'_{ot}}{P'_{or}}\right)^2 \frac{P'_{or}}{P_r}$$

$$= 10 \log_{10} \left(\frac{P'_{ot}}{P'_{or}}\right)^2 + 10 \log_{10} \frac{R_D}{K_L^2} \times \frac{1}{K_d^2 R_b}$$

$$= 2\alpha S + K ,\qquad (9.3)$$

where $\alpha S = 10 \log_{10} \frac{P'_{ot}}{P'_{or}} ,$

the optical fibre loss and K is a constant dependent on the device parameters and the coupling efficiencies.

Thus α = the loss coefficient of the fibre in dB/km and S = length of fibre in km.

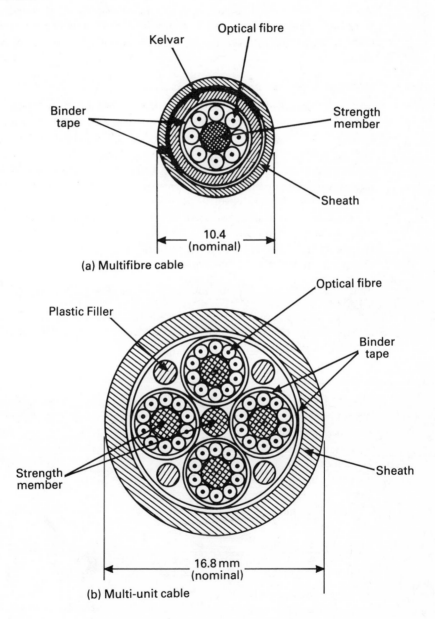

(a) Multifibre cable

(b) Multi-unit cable

Fig. 9.4 – Typical optical fibre cable arrangements.

It will be seen from equation (9.3) that a change of $\delta\alpha$ in optical loss gives rise to a $2\,\delta\alpha$ change in the overall system loss. Hence an increase of x dB in the signal-to-noise ratio of the detector can be obtained by a decrease of $\frac{1}{2}x$ dB in the overall fibre loss. It also follows from equation (9.3) that the electrical signal bandwidth of the overall system will be somewhat smaller than the optical fibre bandwidth. This is because the 3-dB power point for the electrical system loss characteristic corresponds to the 1½-dB power point of the fibre loss characteristic.

OPTICAL FIBRE CABLES

The fragile nature of the single fibre means that they are generally made up into cables. This gives both protection and efficiency in cable-laying operations. Some typical cable structures are given in Fig. 9.4. The relatively small size and high capacity of the optical fibre compared to more conventional transmission media means that it can provide a much higher transmission capacity per cable duct than can be obtained using other means. Table 9.2 gives typical values of the number of telephone channels per duct bore that can potentially be achieved using digital transmission over both conventional metal conductor and optical fibre cables. This capacity can, of course, be utilised for any form of digital transmission. The overall digital transmission rate can be determined by simply multiplying the number of telephone channels by 64 kbits/s.

Table 9.2 – Comparison of optical fibre capacity with conventional transmission systems.

System	Information rate (Mbits/s)	Number of telephone channels	Number of telephone channels per duct bore ($\times 10^6$)
Deloaded junction cable	2.048	30	0.008
Coaxial cable	120	1700	0.07
(40-pair cable)	420	5760	0.23
LED and multimode optical fibre	2.048	30	0.07
	8.448	120	0.27
	34.368	480	1.07
Laser and monomode optical fibre	2.048	30	0.07
	8.448	120	0.27
	34.368	480	1.07
	139.264	1920	4.26
	565	7680	17

10

International communication

Throughout this book we have assumed that most communication takes place over circuits which comprise wire pairs, coaxial cables or optical fibre links and, indeed, most national communication is carried by such media. However, for international communication, especially between countries separated by water, even oceans, it is no simple matter to lay cables. In fact, until recently, almost all international communication, even to the antipodes, was carried by submarine cables. Such cables are high quality coaxial cables, heavily armoured and protected against the ingress of sea water, laid on the ocean bed. The laying of such cables is a very expensive and difficult procedure and maintenance is an almost impossible operation. The cable lengths are such that submerged repeaters are necessary to amplify the multiplexed signals carried by the coaxial bearers. These repeaters have to be super-reliable and must operate at very low power levels since power feed has to be included in the cable structure. Thus the provision and maintenance of sufficient capacity to meet the increasing needs for international communication becomes a major problem. Alternative techniques have thus been developed which avoid the need for submarine cable-laying. However, it will be many years before the submarine cables can be dispensed with entirely. In the mean time they form a vital part of the total telecommunications network.

Even where adjoining countries are not separated by water, they are often separated by rugged terrain and mountain ranges. The inaccessibility of such country and the exposed conditions encountered make it almost impossible to provide cable circuits, either by buried cable or overhead wires. Such conditions are not restricted to international boundaries. There are many places, such as the Alps or even the Scottish Highlands, where such territory is encountered. Even within densely populated areas it is not always a simple matter to obtain way-leave to lay cables along strategic routes.

MICROWAVE RADIO LINKS

The need for cables can be avoided by the use of radio waves. For single speech paths, HF radio can be used. For multiplexed trunk routes the wider bandwidth requirements are more easily met by the use of microwaves. Microwave signals used for communication purposes occupy a frequency band in the range from about 3 GHz to 10 GHz. Perhaps one of the most important properties of micro-wave radio signals is that they propagate rectilinearly, that is, they behave in a way similar to that of light waves. They can thus be propagated as a narrow beam along a line-of-sight path. This conserves transmitter energy by preventing general broadcast scattering of the signals. The signals are launched from para-bolic reflecting aerials situated at geographical high points and received using a similarly sited parabolic receiving aerial. Alternatively, suitable towers can be constructed so that the parabolic aerial 'dishes' can be located above any ob-

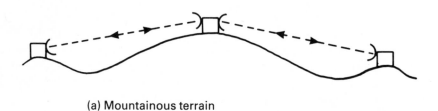

(a) Mountainous terrain

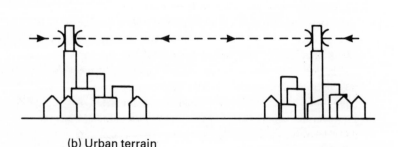

(b) Urban terrain

Fig. 10.1 – Microwave communication links in mountainous and urban environments.

structions likely to be caused by undulating terrain or intermediate high buildings (see Fig. 10.1). Normally microwave links are capable of spanning distances of between 20 and 30 miles, depending on the location of the aerials. Where exceptionally high mountain peaks are available for the location of repeater stations, links in excess of 50 miles are sometimes possible. The first microwave links to be used for telephony were installed in the late 1950s and had a capacity of 240 telephone channels. During the next decade the capacity increased to 1800 telephone channels per radio bearer channel. The telephone signals were assembled using the FDM techniques described in Chapter 4. Thus 1800 telephone channels represents a multiplex of 6 master goups. The FDM signal was then modulated onto the radio frequency carrier using frequency modulation. Consistent with the move towards digital transmission of telephone signals in the rest of the network, it has recently become standard practice for all new systems to operate digitally at 140 Mbits/s, that is, a multiplex of 1920 PCM telephone channels, or an equivalent service. The 140 Mbits/s digit stream is modulated onto the microwave radio carrier using QAM as described earlier in Chapter 6.

Microwave signals are subject to reflection from objects such as hills and buildings. Transmission can also be impaired by weather conditions such as rain and snow. Microwave signals are thus subject to selective fading. It is therefore necessary to incorporate channel quality monitoring and automatic channel switching to ensure continuous satisfactory service.

Despite the high cost of microwave terminal equipment and the need to build expensive towers to mount the aerials, microwave is often cheaper than the cost of negotiating wayleave and the payment of rent for the privilege of access to lay and service an equivalent cable network. In many cases, access makes anything other than a microwave link a practical impossibility.

SATELLITE COMMUNICATION

For intercontinental communication it is obviously impossible to erect microwave towers at 20- or 30-mile intervals across the ocean. However, it is possible to obtain intercontinental communication by a double line-of-sight transmission to communication satellites in orbit around the Earth, as shown in Fig. 10.2. It happens that if a satellite is launched into orbit directly over the equator at a height of 22,000 miles, then its orbital speed is such that, relative to the Earth, it appears to remain, to all intents and purposes, stationary in the sky. Such a satellite is said to be in 'geostationary orbit'. Since signals to and from the satellite pass through the atmosphere at an angle almost normal to the ground, the attenuation of the microwave signal due to atmospheric transmission is minimised. There is thus no problem in transmitting from the ground station to the satellite, and vice versa, in a single hop. The location of the satellite beyond the atmosphere means that it can be powered by solar radiation. It is possible, in principle, to cover almost all the inhabited world with as few as three strategically placed geostationary satellites. In fact many more than this are already

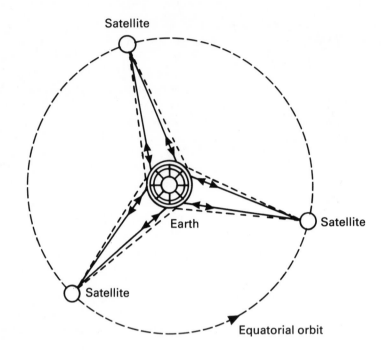

Fig. 10.2 – Communication satellites in geostationary orbit.

in use. Providing the movement of the satellite relative to the Earth is slow enough for it to be tracked by a moveable aerial array, it is not essential that all communication satellites be in geostationary orbit. There are in use a number of satellites which 'rise' and 'set' in the sky on a cyclic basis which can be used for communication purposes for the period of time they are above the horizon to both the sender and the receiver. Satellites are now widely used for inter-continental communication. The one shortcoming is that the length of the communication path can cause annoying delay in the propagation of speech signals. This is especially so where the satellite links have to be used in tandem. The object, therefore, is to provide sufficient satellites to enable, as far as possible, any subscriber anywhere in the inhabited world to communicate with any other using a single satellite communication link.

The satellite contains a pair of microwave transmitters and receivers, one of each pair for each direction of transmission. The signals from Earth are received and detected and then remodulated onto a different carrier for retransmission back to Earth. The power for the satellite equipment can be obtained from solar cells mounted on the surface of the satellite. The somewhat lower power of the return signal means that the ground station receiver has to be of high sensitivity. Again, digital modulation is normally used and forward error correction is often

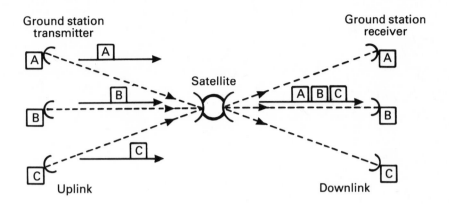

Fig. 10.3 – Basic TDMA operation.

built into the data stream.

Because of the high cost of satellite systems it is essential that efficient use be made of the available bandwidth. More complex multiplexing and modulation techniques are therefore used in satellite systems than in terrestrial microwave links. Early satellite systems used FDM using the standard group arrangements employed in terrestrial communications, later satellites use digital multiplexing. Digital speech interpolation is used. In this technique the silences between words and sentences are utilised for transmission of further signals, which gives an increased utilisation of the transmission capacity. If overload is experienced, a progressive reversion to 7-bit PCM encoding is temporarily implemented until the overload passes. Together, this increases the number of available channels by a factor of 1.9 without any noticeable degradation of the speech quality, even under full loading. Time Division Multiple Access (TDMA) is also widely used on digital satellite links. In this technique several ground stations can each communicate via a single satellite on a time assigned basis. Data is transmitted in bursts from each station in turn, the size of the burst depending on the traffic from each particular station. The signal on the downlink will appear as a single time division multiplexed (TDM) signal and is received in this form by all stations. It is then demultiplexed for onward transmission over the telephone network in the usual way. The basic TDMA operation is illustrated in Fig. 10.3.

11

A postscript — the cellular mobile radio telephone

Since the notes on which this book is based were originally written, there has been a significant step forward in the provision of communication services that it would be imprudent not to mention. This advance is the cellular mobile radio telephone. The purpose of the cellular mobile radio telephone is to provide personal telephone services to subscribers on the move. The greatest demand for the service is likely to be for the 'telephone in the car' by subscribers such as business executives and senior civil servants. The extent of the clientele will be determined by the tariff structure for calls made over the network. There is no reason, however, why the service could not be made available to any mobile subscriber, the communication terminal being carried in the briefcase, or even the pocket. The mobile terminal has normal telephone-type access into the public telephone network, enabling direct telephone communication with any existing telephone subscriber, as well as with other mobile terminals.

Each mobile terminal communicates via a base station using specially allocated UHF channels. The assignment of carrier frequency will be made by the base station. The country is divided geographically into cells as shown diagramatically in Fig. 11.1. Each cell has its own base station. Calls are routed through the base station associated with the cell within which the mobile terminal is located. When a mobile passes from one cell to another, the responsibility for providing the communication channel is handed on and a new carrier frequency is automatically allocated by the base station taking over the responsibility.

To prevent interference between base stations, different frequency allocations are made to neighbouring cells. As can be seen from Fig. 11.1(a), four
different allocations are required if re-use of frequencies is permitted in cells
separated by at least one other cell width. For greater separation a larger number
of allocations is required. This reduces the number of carrier channels available
in each allocation. For a minimum separation of two cell widths, cells have to
be arranged in clusters of seven cells, each with a different frequency allocation,

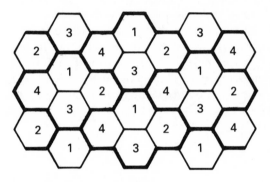

(a) Single-cell separation

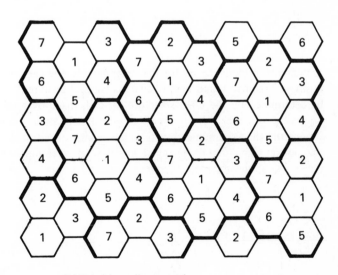

(b) Double-cell separation

Fig. 11.1 — Cellular division of geographical area.

as illustrated in Fig. 11.1(b). The signal powers used are chosen so that satis-factory reception is achieved throughout the whole cell area, but that signals are attenuated so as to become insignificant in cells separated by the minimum re-use distance.

The radio bandwidth allocated for cellular radio telephones is sufficient to provide about 275 separate channel frequency allocations. For seven-call clusters, this gives 39 channels per cell. Using Erlang's formula, the maximum busy-hour traffic to achieve a grade of service no worse than 0.02 with 39 channels is about 15 erlangs. For typical holding times and calling probabilities, 15 erlangs in the busy-hour represents the average traffic generated by about 1500 subscribers. Thus the cell size has to be selected to limit the number of mobiles within the cell area to around 1500. In rural areas, cells of up to 10 miles radius are possible. In urban areas the cell size will have to be reduced. A minimum acceptable cell radius is about 1 mile. If cells are any smaller than this, hand-overs occur too frequently to maintain a stable communication network.

Bibliography

Bear, D. (1976) *Principles of telecommunication traffic engineering.* Peter Peregrinus.

Bennett, G. H. (1983) *PCM and digital transmission.* Marconi Instruments.

Bennett, W. R. and Davey, J. R. (1965) *Data transmission.* McGraw-Hill.

Brown, J. and Glazier, C. V. D. (1974) *Telecommunications,* 2nd Edn. Chapman & Hall.

Bylanski, P. and Ingram, D. G. W. (1976) *Digital transmission systems.* Peter Peregrinus.

Cattermole, K. W. (1969) *Principles of pulse code modulation.* Iliffe.

Carlson, A. B. (1975) *Communication Systems,* 2nd edn. McGraw-Hill.

Cheong, V. E. and Hirschheim, R. A. (1983) *Local area networks.* John Wiley.

Clark, A. P. (1976) *Principles of digital data transmission.* Pentech.

Coates, R. F. W. (1975) *Modern communication systems.* Macmillan.

Davies, D. W. and Barber, D. L. A. (1973) *Communication networks for computers.* John Wiley.

Deasington, R. (1984) *A practical guide to computer communications and networking,* 2nd edn. Ellis Horwood.

Flood, J. E. (ed.), (1975) *Telecommunication networks.* Peter Peregrinus.

Goldman, S. (1953) *Information theory.* Constable.

Griffiths, J. M. (ed.), (1983) *Local telecommunications.* Peter Peregrinus.

Hamming, R. W. (1980) *Coding and information theory.* Prentice-Hall.

Lucky, R. W., Salz, J. and Weldon, E. J. (1968) *Principles of data transmission.*

McGraw-Hill.

Marshall, G. J. (1980) *Principles of digital communications.* McGraw-Hill.

Meadows, R. G. (1976) *Electrical communications.* Macmillan.

National Computing Centre (1982) *Handbook of data communications.*

Richards, D. L. (1973) *Telecommunication by speech.* Butterworths.

Schwartz, M. (1959) *Information transmission, modulation and noise.* McGraw-Hill.

Sherman, K. (1981) *Data communications, a user's guide.* Reston.

Squires, T. L. (1970) *Telecommunications pocket book.* Newnes-Butterworths.

Stallings, W. (1984) *Local networks.* Collier-Macmillan.

Stark, H. and Tuteur, F. B. (1979) *Modern electrical communications.* Prentice-Hall.

Stremler, F. G. (1982) *Introduction to communication systems,* 2md edn. Addison-Wesley.

Taub, H. and Schilling, D. L. (1971) *Principles of communication systems.* McGraw-Hill.

Problems

PROBLEMS: CHAPTER 1

1.1 A carrier is sinusoidally amplitude modulated to a depth of 60% and when this is done the total power transmitted is 12 kW. The depth of modulation is reduced to 40 per cent and it is decided to reduce the total power transmitted to 1 kW by eliminating one side-band and reducing carrier power. By how much will the carrier power be reduced to achieve this objective? Give the answer in decibels.

1.2 Two transmitters, one capable of generating a full AM signal and the other a SSB suppressed carrier signal, have equal mean output power ratings. A single sine-wave modulating signal causes the output from the full AM transmitter to have a modulation index of 0.8. If both transmitters are operating at maximum output, compare, in dB, the power contained in the sidebands of the full AM signal to that contained in the SSB output.

1.3 The instantaneous amplitude of a full AM signal is given by the expression

$$e(t) = 1000 \left[1 + \sum_{n=1}^{3} \frac{1}{2\pi} \cos (2\pi \times 10^3 t) \right] \cos (2\pi \times 10^7 t) \text{ volts.}$$

Determine the peak amplitudes and frequencies of the various components of the modulated signal. Also calculate the mean power that the signal would dissipate in a 500-ohm resistive load.

1.4 Determine the amplitude of the spectral components, up to the sixth side-band pair, of an RF carrier modulated by a 5 kHz sinusoidal signal using a frequency deviation of 25 kHz. Also determine the percentage of power contained in (a) a 10 kHz and (b) 50 kHz bandwidth about the carrier.

PROBLEMS: CHAPTER 2

2.1 A three-stage non-blocking cross-point switching network has 50 inputs and 50 outputs. Calculate the minimum number of cross-points required to implement this network. How many cross-points would be required for a single-stage 50 × 50 cross-point switch? What percentage saving in cross-points can be achieved by using a three-stage network?

2.2 It is proposed to implement a three-stage non-blocking switching network using $n \times n$ switching blocks in each stage. The network has 200 incoming trunks and 200 outgoing trunks. Determine the size of the switch blocks required in each stage to minimise the number of crosspoints used and calculate the total number of cross-points required. Hence show the saving in crosspoint compared to a full 200 × 200 switch matrix.

PROBLEMS: CHAPTER 3

3.1 A full availability group of eight trunks is offered a total traffic of 5 erlangs. Calculate the traffic carried by the third trunk and determine the grade of service of the group.

3.2 A full availability group of ten trunks is offered a total traffic of 4 erlangs. Calculate the traffic carried by each of the first three trunks and the grade of service given by the full ten trunks.

3.3 A traffic load of one erlang is offered to a full availability group of three trunks. The average call duration is two minutes.

(a) What is the average number of calls offered per hour?
(b) What is the probability that no calls are offered during a specified period of two minutes?
(c) What is the proportion of lost traffic?
(d) If the trunks are always tested in the same order, how much traffic is carried by each trunk?

3.4 The arrival rate of telephone calls is two calls per minute. Assuming that the arrivals are independent and equiprobable at all times, find the probability that no calls arrive during any given minute and also find the probability that more than three calls will arrive in 2 minutes. What is the time during which at least four calls will arrive with a probability of more than 95 per cent?

3.5 The traffic offered to a full availability group of six trunks is 3 erlangs. The average call holding time is 3 minutes. What is the average number of calls offered during one hour? Assuming sequential testing of trunks, how much traffic is carried by the first trunk and what is the grade of service? What is the probability of no calls being offered during any specified 3-minute period? (Assume the Poisson distribution for the incidence of calls.)

3.6 Determine the grade of service obtained if an average of 20 calls per hour of average duration 2 minutes are offered to a group of five triunks with full availability. What is the probability of finding no calls in progress at a given time?

3.7 A small company having 4 exchange lines found that, during the busy hour, one in ten attempts to obtain a line resulted in the discovery that all the lines were already engaged. Estimate the busy hour traffic carried by the exchange lines. If a further exchange line is provided, what proportion of abortive attempts to obtain an exchange line would be expected? Determine the grade of service that could be expected with the additional exchange line if the busy hour traffic increases by 50 per cent.

3.8 At a small business establishment the incoming and outgoing calls to the public telephone network are handled by the receptionists, using a PBX with three exchange lines. During the busy hour she handles on average 20 calls per hour, the average call duration being 3 minutes. Calculate:

(a) The traffic load carried by the group of three lines.
(b) The proportion of traffic lost because of non-availability of an exchange line.

If the receptionist leaves the desk for two minutes, what is the probability of at least one call being missed? (Assume that the period of time the caller holds the switchboard before deciding to abort the call is small compared to the receptionist's period of absence).

3.9 During the busy hour a group of six trunks in a local exchange receives calls at a rate of 60 calls per hour. The average call duration is 3 minutes, which gives a grade of service of 0.052. In the first trunk is temporarily taken out of service for maintenance purposes, what would be the grade of service offered during the maintenance period? What will be the increase in traffic temporarily carried by the second trunk in the group?

3.10 A private branch telephone exchange (PABX) serves 100 extensions and has four exchange lines to the public telephone network. During a typical busy hour, an average of one call for every five extensions is made to the public network, the average call duration being 3 minutes. 60 per cent of the total traffic on the exchange lines is the result of the outgoing calls, the remaining 40 per cent resulting from incoming calls from the public network.

Calculate the probability of finding all the exchange lines already engaged when attempting to make a call over the public network from one of the PABX extensions during the busy hour. What is the probability of an outside caller finding all the exchange lines to the PABX busy during the same period?

What would be the improvement in the grade of service to both the inside and outside caller if an additional exchange line were installed?

PROBLEMS: CHAPTER 4

4.1 A transmission line has the following parameters per loop kilometre:
$R = 50 \ \Omega$, $L = 0.5$ mH, $C = 0.005 \ \mu$F and $G = 10^{-4}$ S. What is the characteristic impedance at 100 Hz and 1 MHz?

By how much must the inductance be changed to achieve distortionless transmission?

4.2 An audio cable has primary coefficients per loop kilometres of $L = 0.6$ mH, $C = 0.05 \ \mu$F, $R = 70 \ \Omega$. Leakage is assumed negligible. What is the characteristic impedance of this cable at 4 kHz? If the cable is now loaded with 88 mH every 1.8 km, how does the characteristic impedance change in magnitude and angle?

4.3 A high frequency transmission line having a characteristic impedance of 75 Ω is used as a feeder to an aerial. The aerial input impedance is 150 $+ j150 \ \Omega$ at the frequency being used. What is the standing wave ratio on the line and the reflection coefficient at the aerial terminals?

4.4 A low-loss transmission line of characteristic impedance 75 Ω is used to feed signals to a terminating load. A slotted-line test bed is used to determine the voltage standing-wave ratio and the distance of a voltage minimum from the load termination at two frequencies. The results obtained are as follows:

Frequency	VSWR	Distance of V_{min} from load
100 MHz	3.1	50.7 cm
300 MHz	5.2	31.6 cm

Determine from this data an equivalent circuit for the terminating load. Assume the velocity of phase propagation in the transmission line and the slotted line is 3×10^8 m/s.

4.5 Calculate the overall noise factor and the effective noise temperature of a three-stage amplifier where the parameters of the constituent stages are as follows: Stage 1, Noise temperature $T_1 = 4$ K, gain $G_1 = 30$ dB; Stage 2, Noise factor $F_2 = 6$ dB, gain $G_2 = 20$ dB; Stage 3, Noise factor $F_3 = 12$ dB, gain $G_3 = 60$ dB. Assume $T_0 = 290$ K.

4.6 The input impedance of an amplifier is 20 k Ω and is fed from a matched source. Noise produced within the amplifier may be represented by an effective increase in temperature of the source impedance of 439 K. What is the noise factor, in decibels, of this arrangement?

4.7 A receiver for geostationary satellite transmissions at 2 GHz consists of an antenna pre-amplifier with a noise temperature of 124 K and a gain of 20 dB. This is followed by an amplifier with a noise figure of 12 dB and a gain of 80 dB. Calculate the overall noise figure and equivalent noise temperature of the receiver.

4.8 The receiver in the previous question has a bandwidth of 1 MHz. The receiving antenna gain is 40 dB and the antenna noise temperature is 59 K. If the satellite antenna gain is 6 dB and expected path losses are 180 dB, what is the minimum required satellite transmitter power to achieve a 14 dB S/N ratio at the output of the receiver?

PROBLEMS: CHAPTER 5

5.1 Devise a low-redundancy binary code for the following message set:

Message	Probability of occurrence
1	0.5
2	0.3
3	0.1
4	0.05
5	0.03
6	0.02

Calculate source entropy and hence determine the code efficiency.

5.2 A message source selects messages from a set of eight messages having the following *a priori* probabilities of occurrence:

Message	Probability of occurrence
1	0.25
2	0.25
3	0.2
4	0.15
5	0.08
6	0.04
7	0.02
8	0.01

Devise a low-redundancy binary code which may be used to transmit messages from this source and calculate its efficiency. How does this compare with straight binary encoding using 3 bits per message?

5.3 Calculate the source entropy, equivocation, mutual information and channel capacity for a binary symmetric channel in which $p(x_1) = p(x_2) = 0.5$ and $p(y_n/x_n) = 0.99$ $(n = 1,2)$.

5.4 A ternary channel has forward probabilities $p(y/x)$ as follows:

	Output messages		
	y_1	y_2	y_3
x_1	0.988	0.01	0.002
x_2	0.01	0.98	0.01
x_3	0.002	0.01	0.988

Input messages

The *a priori* message probabilities are:

$$p(x_1) = p(x_3) = 0.25,$$
$$p(x_2) \qquad\quad = 0.5.$$

Calculate the source entropy, the equivocation and the mutual information. What proportion of the received symbols will be in error?

5.5 A transmission channel has forward probabilities as follows:

	Output messages			
	y_1	y_2	y_3	y_4
x_1	0.9	0.025	0.025	0.05
x_2	0.1	0.8	0.05	0.05
x_3	0.1	0.1	0.7	0.1
x_4	0.1	0.1	0.2	0.6

Input messages

The *a priori* message probabilities are

$$p(x_1) = 0.2, \ p(x_2) = 0.3, \ p(x_3) = 0.4, \ p(x_4) = 0.1.$$

Calculate the source entropy, the equivocation and the mutual information for the channel. Determine the maximum value of mutual information that could be obtained by statistical matching.

5.6 A signal with a power spectral density given by

$$s(f) = \frac{S(0)}{1 + \left(\dfrac{f}{f_c}\right)^2}$$

is sampled at a rate of f_s. What must be the ratio of the sampling frequency f_s to the spectrum half-power frequency f_c if the signal-to-aliased power ratio is to be at least 26 dB?

5.7 A waveform consisting of a fundamental sinusoid of frequency 500 Hz and a third harmonic of amplitude 0.1 times that of the fundamental is sampled at a rate of 10 ksamples/s by a pulse having a duration of 20 μs. Calculate the frequencies and relative amplitudes of the first ten components of the spectrum of the signal output from the sampler.

5.8 The error-rate in the received signal when binary data is transmitted through a transmission channel with gaussian noise is 1 in 10^4. Select a Hamming forward error-correcting code which will reduce the error rate to 1 in 10^6. State how much information redundancy is needed to implement this code.

5.9 A data transmission link is available which will convey digital information at a rate of 4800 bits/s with an error-rate of 1 in 10^4, the errors occurring at random. Select a Hamming forward error-correcting code which will reduce the bit error-rate to about 1 in 10^7. Determine the average rate of information transmission (data throughput) that can be achieved using this code for transmitting data over the link.

5.10 A binary data transmission system operates with data blocks consisting at 15 bits each. The bits within the block may be either data-bearing bits or check bits for error detection or correction purposes. If the error-rate before correction is 1 in 10^3, estimate the number of bits, on average, that will need to be transmitted in order to transmit 10^6 information bits with all single errors connected both for ARQ and forward error-correcting codes. Also estimate the number of bits still in error after the correction of all single errors.

PROBLEMS: CHAPTER 6

6.1 A series of rectangular pulses of 0.01 ms duration are transmitted at intervals of 0.1 ms. Determine the relative amplitudes of the first four components of the transmitted signal. What pulse duration would be required in order to make the fourth component equal to zero?

6.2 A bipolar binary signal is passed through a channel which can be represented by a simple RC low-pass filter with a 3 dB frequency f_c. The bit duration of the binary signal is $T = 1/f_c$. Sketch the eye patterns for the following signal sequences:

(a) An alternating 1010 sequence.
(b) An alternating 11001100 sequence.
(c) An alternating 1111101111 10 sequence.

Can you tell from the eye patterns where the probability of error is greatest?

6.3 A train of random binary bipolar data pulses is transmitted at a rate of 600 bauds. What is the minimum bandwidth required to transmit the data? If a signal consisting of alternate pairs of positive and negative pulses (i.e. 1, 1, 0, 0, 1, 1, 0, 0, etc.) is passed through an ideal filter of minimum bandwidth, what is the shape of the received signal?

6.4 A data communication system transmits data using four-level amplitude-modulated rectangular pulses of duration 100 μs at a pulse-rate of 10 kpulses/s. The four amplitude levels are equally spaced and are symmetrical about zero ·volts. The bandwidth of the transmission path is sufficiently large to avoid significant distortion of the pulse shape at the receiver. The system conveys random binary data encoded into the four levels by one of the two following codes:

Level	Code A	Code B
+3	11	11
+1	10	01
−1	01	00
−3	00	01

The received signal is impaired by additive gaussian noise, the signal-to-noise ratio at the receiver being 20 dB. Estimate the bit-error rates obtained for the coding schemes A and B respectively.

6.5 Assuming random data and a channel of sufficient bandwidth to avoid significant distortion of the pulse shape in transmission:

    (a) estimate the error-rate for a bipolar binary and an AMI line coded transmission system for signal-to-noise ratios of 10 dB, 13 dB and 15 dB.

    (b) estimate the minimum recieved signal-to-noise ratio for both bipolar binary and AMI line coded transmission systems in order that the average bit-error-rate shall not exceed 1 in 10^5.

6.6 Binary digits are transmitted using a pseudo-ternary Alternate Mark Inversion line code. The binary digits generated by the source are random, that is $P_0 = P_1 = 0.5$, where P_0 is the probability of binary 0 and P_1 is the probability of binary 1. The transmitted signal is impaired by zero-mean white gaussian noise such that at the receiver the signal-to-noise ratio is 12.5 dB. Calculate:

    (a) The probability of error in the received signal.

    (b) The equivocation of the received signal.

    (c) The efficiency of the line code used.

The probability of the transmitted positive mark condition being received as the negative mark condition or vice-versa can be neglected.

PROBLEMS: CHAPTER 7

7.1 How many code bits per sample are required in a linearly quantised PCM system if the quantisation noise power is to be at least 40 dB less than the power of a sinusoidal signal which just occupies the full quantisation range without distortion due to overload?

7.2 A 7-bit PCM system employing uniform quantisation has an overall output digit rate of 56 kbits/s. Calculate the signal-to-quantising noise ratio that would result when the input is a sine wave with peak-to-peak amplitude equal to the available amplitude range of the system. Calculate the dynamic range for sine wave inputs in order that the signal-to-quantising noise ratio may not be less than 30 dB. What is the theoretical maximum sine wave frequency that the system can handle?

7.3 Three 5 MHz band-limited signals are to be transmitted using PCM. Calculate the maximum sampling period and the maximum time the signal sampler can dwell on each channel. If the system is to be designed to have a signal-to-quantising noise power ratio of at least 50 dB, determine the minimum number of quantisation levels required and the gross bit rate of the complete system, assuming one bit per sample is also transmitted for signalling and synchronisation purposes.

7.4 A linearly quantised PCM system is required to give the same signal-to-quantisation noise ratio as A-law companded quantisation with 8 bits per sample when the signal is 40 dB below the maximum signal amplitude that just occupies the whole of the quantisation range without distortion. How many bits per sample does the linear system require?

7.5 Assuming a sinusoidal signal occupying the full quantisation range of the system, calculate the signal-to-quantisation noise ratio:

(a) with CEPT 30-channel PCM companding law,
(b) with equally spaced quantisation steps and the same total number of bits.

If the amplitude of the input signal is reduced by 21 dB, what will be the signal-to-quantisation noise ratios for both schemes above?

PROBLEMS: CHAPTER 9

9.1 A 5-km multimode optical fibre link has an optical loss coefficient of 3 dB/km. If the fibre is subsequently replaced with a new fibre having an optical loss coefficient of 2 dB/km, what increase in electrical signal power can be expected at the photodiode detector output?

9.2 The fibre used in a multimode optical fibre link transmitting signals over a distance of 2 km has an optical loss coefficient of 3 dB/km. This is replaced by a new fibre having an optical loss coefficient of 1.5 dB/km. Assuming the transmitted signal remains unchanged, what improvement in signal-to-noise ratio at the photodiode detector output can be expected as a result of this improvement?

Appendix:
The determination of error-rate
in the presence of noise

The effect of noise in a digital transmission system is to produce errors in the receiver signal. The actual probability of error for a given signal-to-noise ratio is dependent on the nature of the noise, the way in which the data is modulated onto the carrier and the way in which the decision thresholds are set in the signal detector. A comprehensive analysis would require the consideration of a number of different factors. In this appendix we shall restrict our analysis to bipolar binary and multilevel pulse amplitude modulation systems with a simple decision threshold set halfway between the voltage levels of adjacent permissible modulation levels in the received signal. We shall also assume that the noise has a gaussian amplitude distribution. This condition gives the worst-case probability of error for a given signal-to-noise ratio, any departure from the gaussian distribution reducing the entropy of the noise signal. The gaussian noise voltage distribution is given by

$$P(V) = (2 \pi \sigma^2)^{-\frac{1}{2}} \exp(-V^2/2 \sigma^2) ,$$

where σ^2 is the noise variance and represents the mean power of the noise signal. The guassian distribution curve is given in Fig. A.1.

The probability that the voltage V exceeds x at a given instant is given by the area shown shaded in Fig. A.1. This area we shall define as

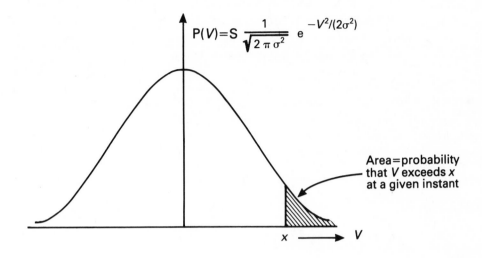

Fig. A.1 – Gaussian distribution.

$$\mathrm{Erfc}(x) = \int_x^\infty (2\,\pi\,\sigma^2)^{-\frac{1}{2}} \, \exp\left(-V^2/2\,\sigma^2\right) \mathrm{d}V \ . \tag{A.1}$$

It is usual to measure x in units of σ rather than in volts, whence we can normalise equation (A.1) so that

$$\mathrm{Erfc}(x) = \int_x^\infty (2\,\pi)^{-\frac{1}{2}} \, \exp(-Z^2/2) \, \mathrm{d}Z \ ,$$

where $Z = V/\sigma$.

Communications engineers often refer to $\mathrm{Erfc}(x)$ as the error-function and tables of $\mathrm{Erfc}(x)$ are sometimes available. However, the error-function of mathematics and statistics, which is usually denoted $\mathrm{erfc}(x)$ (note the lower-case 'e'), is related to $\mathrm{Erfc}(x)$ such that

$$\mathrm{Erfc}(x) = \tfrac{1}{2}\,\mathrm{erfc}(x/\sqrt{2}) \ .$$

The function $\mathrm{erfc}(x)$ is to be found in most books of mathematical tables. Care must be taken to ensure the two related functions are not confused in carrying out error-rate calculations.

ERROR-RATE FOR A BIPOLAR BINARY SIGNAL IMPAIRED BY GAUSSIAN NOISE

For a bipolar binary signal, where the 0 and 1 binary states are represented by voltage amplitudes of $-V$ and $+V$ respectively, the mean signal power is V^2. The mean noise power is σ^2, thus the signal-to-noise ratio S/N is $V^2/\sigma^2 = (V/\sigma)^2$. The error threshold is at zero volts, thus the probability of error is given by the area of the tail of the gaussian distribution where $x = V/\sigma$ in normalised amplitude units. This is so whether the transmitted data is 0 or 1.

Thus the probability of error

$$
\begin{aligned}
P(e) &= \mathrm{Erfc}(V/\sigma) \\
     &= \mathrm{Erfc}(S/N)^{\frac{1}{2}} \\
     &= \tfrac{1}{2}\,\mathrm{Erfc}(S/2N)^{\frac{1}{2}} \; .
\end{aligned}
$$

(A.2)

ERROR RATE FOR AN L LEVEL BIPOLAR AMPLITUDE MODULATED SIGNAL IMPAIRED BY GAUSSIAN NOISE

From Chapter 5 we note that the mean signal power for an L level signal is given by

$$
S = K^2 \sigma^2 (L^2 - 1)/12 \; ,
$$

where the levels are spaced apart at voltage intervals of $K\sigma$ volts. Hence the signal-to-noise ratio

$$
S/N = K^2(L^2 - 1)/12
$$

(A.3)

and the error threshold is equal to $(K\sigma)/2$ volts, that is $K/2$ in normalised units.

For levels 2 to $L - 1$, the probability of error is given by the area in both tails of the gaussian distribution, for errors can result from either positive or negative perturbations of the amplitude level. For levels 1 and L, the probability is given by the area of a single tail only. Assuming all levels are equiprobable, the overall probability of error is given by

$$
\begin{aligned}
P(e) &= \{2(L-2) + 2\}/L\;\mathrm{Erfc}(K/2) \\
     &= 2\,(L-1)/L\;\mathrm{Erfc}\,(K/2).
\end{aligned}
$$

From (A.3) $K = \{12\,(S/N)/(L^2 - 1)\}^{\frac{1}{2}}$

Thus $\begin{aligned}
P(e) &= 2\,(L-1)/L\;\mathrm{Erfc}\,\{3(S/N)/(L^2-1)\}^{\frac{1}{2}} \\
     &= (L-1)/L\;\mathrm{erfc}\,\{1.5\,(S/N)/(L^2-1)\}^{\frac{1}{2}} \; .
\end{aligned}$

Note that for the bipolar binary case $L = 2$, which gives, by substitution,

$$P(e) = \tfrac{1}{2}\,\text{erfc}(S/2N)^{\tfrac{1}{2}} \ .$$

This confirms the result derived earlier and given in equation (A.2).

Answers to problems

1.1 9.17 kW or 90.2 per cent.

1.2 −6.2 dB.

1.3 1000 V, 10^7 Hz; 250 V, $(10^7 \pm 10^3)$ Hz; 125 V, $(10^7 \pm 2^6)$ Hz; 83 V, $(10^7 \pm 3 \times 10^3)$ Hz; 11.7 kW.

1.4 0.18, 0.33, 0.05, 0.36, 0.39, 0.26, 0.13.
     (a) 24.6 per cent. (b) 95.9 per cent.

2.1 1800, 2500, 28 per cent.

2.2 Primary and tertiary each 20 off 10×10, secondary 10 off 20×20, 8000;
     200×200 requires 40,000, saving is 80 per cent.

3.1 0.73 E, 0.07.

3.2 0.8 E, 0.74 E, 0.655 E, 0.0053.

3.3 30, 0.368, 0.0625, 0.5 E, 0.3 E, 0.1375 E.

3.4 0.1353, 0.5665, 3.9 minutes.

3.5 60, 0.75 E, 0.058, 0.05.

3.6 0.00057, 0.514.

3.7 1.8 E, 1 in 27, 0.11.

3.8 0.9375 E, 0.0625, 0.4866 (i.e nearly 50 per cent).

3.9 0.11, 13 per cent.

3.10 0.0624 (same for internal and external caller); 0.0203 (i.e. by a factor of about 3).

4.1 707 Ω, 301 Ω, +2 mH/ loop km.

4.2　$238.7\ \Omega\angle 38.92°$, changes to 995.6 ohms $\angle 1.61°$.

4.3　$4.26, 0.62\angle 30°$.

4.4　Series circuit consisting of $R = 75\ \Omega, L = 0.1$ H, $C = 10$ pF.

4.5　1.01695, 4.9135 K.

4.6　4 dB.

4.7　1.98 dB, 167.3 K.

4.8　49.5 kW.

5.1　E.g. 0, 10, 110, 1110, 11110, 11111. 1.834 bit/message, 99.1 per cent.

5.2　E.g. 00, 01, 10, 110, 1110, 11110, 111110, 111111.
99.29 per cent, 84.39 per cent (a 15 percent saving in bits).

5.3　1 bit/message, 0.0808 bit/message, 0.9192 bit/message, 0.9192 bit/message.

5.4　1.5 bit/symbol, 0.127 bit/symbol, 1.373 bit/symbol, 1 in 62.5.

5.5　1.842 bit/symbol, 1.051 bit/symbol, 0.791 bit/symbol, 0.89 bit/symbol.

5.6　509.2.

5.7　500 Hz, 1.0;　1500 Hz, 0.1;　8500 Hz, 0.094;
9500 Hz, 0.935;　10,500 Hz, 0.935;　11,500 Hz, 0.094;
18,500 Hz, 0.076;　19,500 Hz, 0.757;　20,500 Hz, 0.757;
21,500 Hz, 0.076.

5.8　(67,61) or (63,58); 91 per cent or 92 per cent.

5.9　(11,7), 3000 bits/s.

5.10　1,088,190, 1,363,500, 15, 15.

6.1　1.0, 0.984, 0.935, 0.858, 0.033 ms.

6.2　The probability of error is greatest after a long run of the same binary symbol.

6.3　300 Hz, 150 Hz sinusoid.

6.4　8 in 10^6, 6 in 10^6.

6.5　(a) 1 in 1250, 1 in 250,000, 1 in 10^8; 1 in 50, 1 in 833, 1 in 20,000.
(b) 12.6 dB, 15.8 dB.

6.6　0.00216, 0.0302, 63 per cent.

7.1　7.

7.2　43.8 dB, 13.8 dB, 4 kHz.

7.3　0.1 μs, 33.3 ns, 512, 300 Mbits/s.

7.5　37.9 dB, 50 dB, 37.9 dB, 13.8 dB.

9.1　10 dB.

9.2　6 dB.

Index